Inside the
Animal World

Inside the Animal World

An Encyclopedia of Animal Behavior

Maurice and Robert Burton

NYT

Quadrangle/The New York Times Book Co.

Title page: The Hamadryas, Arabian or Sacred baboon, which lives in areas on either side of the Red Sea, spends the day foraging in small groups, turning over stones for insects and scorpions. The dominant male of this party is giving an aggressive signal, warning members of an approaching party to keep their distance.

Color photography by Jane Burton
Photographs supplied by Bruce Coleman Limited
Line illustrations by Hilary Burn

First published in Great Britain in 1977 by Macmillan London Limited

Library of Congress Catalog Card Number: 76-52819
International Standard Book Number: 0-8129-0688-8

Contents

Chapter One
The Springs of Behaviour

introduction: ethology: European and American schools of ethology — reflex behaviour: stimulus and response: attenuation — kinesis: taxis — instinctive and innate behaviour: motivation: drive: appetitive behaviour: satiation: conflict situation: displacement behaviour — role of hormones — activity cycles: biological clock: circadian cycles — fixed action pattern — learning: habituation: association: conditioning: trial and error: latent learning: insight: problem-solving — evolution of behaviour — intelligence

From the beginning, before the first glimmerings of civilization, man has studied animal behaviour. It has been an essential part of his struggle for survival. Our remote ancestors lived by hunting and gathering; hunting larger animals and gathering insects and shellfish, as well as berries, nuts and roots. The success of this way of life must have lain in acquiring a knowledge, often intimate, of the habits of many animals. Primitive man had to know where he was most likely to find particular animals, and in what season. He would have learned that ants swarmed on warm, still days and that salmon spawned in shallow streams in autumn, and he would have adjusted his life accordingly. To catch bigger game, hunters had to know even more intimately the private lives of animals. Their spears and arrows were only effective over a range of about twenty yards, so they had to get close to prey that was aware that man was a predator. To achieve this, the hunters had to make themselves familiar with the habits of their quarry, its tracks through the forests, its waterholes, its favourite foods and whether it would stand its ground to defend its young.

An idea of what the hunters learned about their quarry is

Impala, small African antelopes, at a waterhole. Animals, such as lions, as well as primitive hunters, quickly learn that energy can be saved and a more certain prospect of a kill ensured, by ambushing thirsty prey when they come to drink.

16

given by D.R. McCullough who studied elk (the North American equivalent of the European red deer) at close quarters. He could easily walk to within two hundred yards of the elk and could get much closer if he slowed his movements. The important point was that the elk could readily detect movements across its field of vision. So he used the slow, fluid movements of a cat stalking a bird and, moreover, if he made a frontal approach his apparent rate of movement was less than if he had crossed the deer's field of vision. Physiologists tell us that the eye is very good at detecting movement because light from a moving object stimulates each sense cell or group of sense cells in turn as it traverses the retina. Early hunters were unaware of this nugget of information, but they must have been as aware of its implication as the modern zoologist.

There is no doubt that primitive hunters knew as much about many of the practical aspects of animal behaviour as any man since. We can see this today among peoples that still live by hunting or gathering, such as the Eskimos of the north and the Kalahari bushmen. The Athabascan Indians of Canada, for example, had an intimate relationship with the herds of caribou, which supplied all their needs. The caribou's hide was used for clothing, sleeping bags and covers for canoes, its fat for lighting and the antlers for a number of tools and utensils. The Indians' way of life depended on the migrations of the herds.

With the development of agriculture, people, particularly the richer and more civilized, lost their close relationship with wild animals — until some thousands of years later inquisitive men started to study their habits, not out of necessity but for pleasure. These first naturalists, at first content to describe what they saw, were soon led to ask questions about why animals behave as they do and what were the mechanisms of behaviour. So the science of animal behaviour came into being.

The study of animal behaviour is new as a science, being very much the junior of anatomy and physiology, but its foundations lie in the very precise observations of the old naturalists and, as careful recording is still its essence, it can be said to have started with the work of J.H. Fabre, who died in 1915.

Fabre spent more than forty years closely observing the ways of insects, and moreover he watched them in their natural surroundings, in his garden at Serignan in the south of France and elsewhere. Apart from giving us delightful descriptions of the lives of crickets, beetles and moths, Fabre speculated on the driving force of animal behaviour. He considered that every animal had a life-plan that guided its actions. Its behaviour was said to be instinctive, and instinct was thought to account for most of its actions, guiding the animal so as to preserve both its own life and the life of its species.

An important step forward in the study of animal behaviour came with the realization that descriptions of behaviour must be coldly factual. Darwin and Fabre related the actions of animals in terms of human feelings. Darwin described the expression of a chimpanzee as 'disappointed and sulky', which are human emotions, so he was guilty of anthropomorphism — treating an animal as if it were a kind of human being. It is not easy to avoid the anthropomorphic pitfall, particularly when it helps to give a picture of what the animal is doing. The chimpanzee *does* look disappointed and sulky but to say so invites a completely false interpretation of its behaviour and credits an animal with greater intelligence than it really possesses.

The necessity for a critically analytical approach to animal behaviour is neatly shown by the story of the performing horse 'Clever Hans', who astonished everyone with his ability to count by tapping a hoof. He could tell the date and time as well. Eventually a psychologist investigated the phenomenon and found that instead of a high intelligence, Clever Hans had a keen eye. As the trainer set the question he would involuntarily lean forward a fraction and Clever Hans would tap away until he noted the slight and still involuntary jerk of the trainer's head which showed that he had reached the right answer. If Clever Hans could not see the trainer he was immediately revealed as an imposter. The moral of the story is that things are not always what they seem and that it is unwise to accept an extravagant theory when a simpler explanation is possible.

The scientific study of behaviour, or **ethology**, deals with a wide range of topics, from the simplest automatic reactions of microscopic organisms to the more familiar and widescale actions of apes, our nearest relatives. It includes much of human conduct and reactions, although human behaviour is usually treated as a separate subject, because it is immensely complex and because of its great importance to all of us.

As the study of behaviour is in its infancy, theories concerning the mechanisms of behaviour are continually

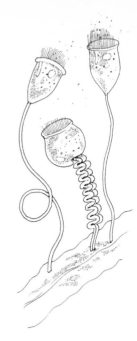

Crying wolf!
An animal must be continually on guard against danger. If its vigilance relaxes it soon falls prey to the sudden onslaught of a predator. Yet if it bolts at every slight alarm it will never be able to go about its daily business. If a potentially harmful stimulus is repeated at intervals, the animal becomes **habituated** to it and its response to it becomes **attenuated**.
In the mud of river estuaries lives a ragworm *Nereis,* a distant relative of earthworms. It lives in a burrow and its head end emerges while it feeds on plant and animal debris. In this position it is vulnerable to attack by carnivorous fishes, so the slightest hint of alarm — a vibration or a passing shadow — causes an immediate retreat. About a minute later the worm slowly emerges again. After repeated stimuli, a series of shadows perhaps, the ragworm

18

stops retreating. The shorter the interval between shadows, the sooner the worm becomes habituated. A series of shadows is more likely to come from a frond of seaweed waving in the swell, or a stream of flotsam floating on the tide, than from a predatory fish; so the ragworm is adapting to its environment. When the stimulus ceases the ragworm loses its habituation in about twenty-four hours; but while habituated to shadows it will still react to other stimuli.

Vorticella is a freshwater protozoan whose bell-shaped body is attached to the substratum by a slender stalk. A slight jar causes the stalk to coil up like a bed spring, drawing the body out of harm's way. Repeated jars cause habituation, a behavioural trait that is very useful for *Vorticella* that are attached to Pond snails, tadpoles and other small animals and are therefore continually on the move.

changing. A theory is set up and facts are collected or experiments performed to see whether they fit it. If not, the theory has to be altered or abandoned. Soon after animal behaviour study started, the investigators proposed theories or models of behaviour to explain what they had observed and to guide their future research. The original theories had the advantage of simplicity but to zoologists they now appear naive. This is a pity for anyone who is trying to get to grips with a difficult subject. The early models of animal behaviour were relatively easy to understand, whereas modern theories have become harder to appreciate. They tend, for instance, to employ analogies from computer technology which is itself incomprehensible to many, and as they become more sophisticated they necessarily become more complex. Nevertheless, it is still possible, without a specialist background, to grasp sufficient of the elements of animal behaviour to be able to appreciate and improve one's enjoyment of watching animals.

Reflex behaviour

By studying insects and other lowly animals, naturalists came to view all animals as mere machines and their behaviour as the resulting product of a series of automatic reactions. These naturalists were known as mechanists and they were supported by the thinking of the French philosopher Descartes.

The simplest form of behaviour, animal or human, is the **reflex.** It is automatic and, by and large, does not vary. For example, the knee-jerk reflex, used by doctors to test certain patients, is completely automatic. We have no control over it; every time the doctor taps the tendon below the knee cap, the foot shoots out. There is a very simple mechanism, consisting of a **stimulus** (a tap with a hammer in this case) and a **response** (the foot jerking out).

In the knee-jerk reflex, the stretching of the tendon, as happens when it is tapped, pulls a muscle in the knee and excites certain cells, the stretch sensors, among the muscle fibres. The stretch sensors send impulses along nerve fibres to the spinal cord. There they stimulate the production of impulses along another set of nerve fibres that return to the muscle, stimulating it to contract sharply. The course taken by the impulses is called a nervous pathway, and the kind described here is called a reflex arc.

Although reflex behaviour appears to be of a rigid pattern it is capable of being modified to some extent. Thus, when a dog has its flank rubbed it automatically brings up its hind leg in a scratching action. After twenty seconds of continuous rubbing, however, the leg movement slows down and finally stops. This happens not because the dog has become bored with scratching nor because its leg is getting tired but through a process of adaptation, a change in the nervous mechanism of the reflex arc. In fact, if the rubbing movements are moved a few inches along the flank, the scratching will start up again.

Another reflex that can be easily investigated is the ear reflex of cats. Repeated tickling of a cat's ear causes it to be laid back and fluttered rapidly. After continued tickling, the cat shakes its head and finally it scratches the ear with a hind leg. Viewed uncritically, the sequence of actions looks like a deliberate attempt to remove a source of irritation but the cat has no say in the matter. The movements are reflex and as automatic as a knee-jerk, but they are clearly more complex. The ear reflex is more of a collection of reflexes so closely related that they cannot be separated. They appear in order because each requires a different level of stimulation. Little stimulation is required to make the ear lie back but continued stimulation is needed to cause scratching of the ear with the hind leg.

Another way that reflex behaviour can be modified is by **conditioning,** in which the animal comes to respond to a particular stimulus with an action appropriate to another stimulus. This is called a conditioned reflex. It is a form of learning and will be explained in detail later.

Reflexes form the oldest kind of behaviour and are found throughout the animal kingdom. They are often concerned with avoidance of danger, as, for example, withdrawing the bare foot from a thorn. Automatic, stereotyped and never-varying patterns of action are an advantage in such circumstances. But even in the single-celled protozoans more complex forms of behaviour can be seen. They are like reflexes in that they are automatic and involve simple nervous mechanisms, but they also serve to guide the animal towards or away from part of its environment and involve movement of the whole animal.

Kinesis and taxis

A **kinesis** is a kind of behaviour in which the animal is not guided in the strict sense. It is stimulated into action but wanders about aimlessly. Its behaviour is nevertheless being controlled. In an **orthokinesis,** stimulation changes the speed of movement. A rise in humidity stimulates woodlice to slow

Vibrations guide the hunt
The backswimmer or Greater waterboatman *Notonecta,* a familiar bug of ponds and lakes, habitually floats with its boat-shaped body upside down and propels itself by rowing with its long, oar-like hindlegs. It feeds by piercing the bodies of small animals with its rostrum or beak and sucking their juices. Its main prey is insects that fall onto the surface of the water and lie there buzzing, unable to escape. The backswimmer rests with the tips of its first two pairs of legs touching the surface of the water and vibrations from the trapped insect trigger the backswimmer's **appetitive behaviour** — (see p 25) the mechanism whereby it searches for its food.

There is a set sequence of events. For the backswimmer to be alerted, its prey must set up vibrations in water at frequencies between 5 and 500 cycles per second. On perceiving them, the backswimmer turns to face the source then swims towards it. It can detect the vibrations from a range of six inches, then, when an inch away, it is guided to the target by sight. Before being pierced with the rostrum, the quarry is tested with the mouthparts to make sure that it is edible.

down whereas in dry air they run about actively. This can be
tested experimentally by putting woodlice in a glass-topped
vessel with a false floor of gauze. A dish of water placed
under the floor keeping the air humid, causes the woodlice to
move about slowly or stop altogether. But if the water is
replaced with concentrated sulphuric acid that mops up
water vapour from the air, the woodlice move about more
rapidly. Simple as this behaviour is, it is essential for the
survival of the woodlice. The hard skin, or cuticle, of a
woodlouse is not completely waterproof so that in dry air its
body will lose water by evaporation. The orthokinesis causes
the woodlice to gather in groups in damp places such as
under stones or logs.

A **klinokinesis** is behaviour characterized by a change in
rate of turning. It is a more complex pattern than an
orthokinesis, as the animal moves in a straight line while in a
favourable environment but on entering an unfavourable
environment it starts to wheel about, as if trying to find its
way out. However, the turns are at random and it is only
chance that leads the animal out to better conditions. Once
there, its course straightens out and it tends to move away
from danger. The human body louse lives in the hair or
between clothes and the skin where it is warm and humid. If
dislodged, it starts to crawl hither and thither, again looking
as if it is purposely seeking a better place but, in fact, turning
at random until it happens to encounter more favourable
conditions.

A **taxis** is behaviour in which an animal steers with respect
to a source of stimulation, such as a male seeking a female or
a moth flying into a light. Steering is effected by measuring
the strength of the stimulation and moving along a gradient
of stimulation either towards the source **(positive taxis)** or
away from it **(negative taxis)**. The simplest form of tactic
behaviour, **klinotaxis**, is shown by mealworms which have a
negative **phototaxis** that leads them to escape from light by
crawling into crevices. The mealworm compares the intensity
of the light on each side of the body in turn and orientates
itself so that it will receive equal stimulation on each side.
When this is achieved, it will be facing towards or away from
the source of stimulation. A more advanced form is
tropotaxis in which the stimuli on each side are compared
simultaneously by paired sense organs such as the eyes or
ears. An adult fly can move straight towards a light in this
fashion but, if one eye is covered with a drop of paint, it can
only fly in circles as it forlornly attempts to balance the

21

stimulation. A honeybee, however, can still fly towards a light when one eye is covered. This is called a **telotaxis.** The bee can compare the strength of light falling on different parts of the eye.

Instinct and learning

Fabre and the mechanists took too simple a view of animal behaviour. At even the lowest level animals are something more than machines. They do not always react to a particular situation in precisely the same way. To take a human analogy, our reaction to the unexpected visit of a friend is not always the same. Although usually welcome, there are times when a visitor is not appreciated. It depends on our **mood,** and mood is a term used to denote the internal state of an animal that leads it to behave in one way or another. When an animal changes its behaviour for no apparent reason, we can say that its mood or **motivation** has changed. Something, either inside its brain or affecting the brain, has altered and one of the quests of the behaviour student is to find what is happening within the brain to control the visible behaviour. Not surprisingly, this has proved a difficult task as the brain cannot be dissected to show how it operates in the same way as a hand or leg can be taken apart to show the operation of bones, muscles and ligaments.

In the quest for understanding of the mainsprings of behaviour, two rival schools of study developed. In Europe, Konrad Lorenz established a following who considered that much of animal behaviour is inherited from generation to generation. It is readily observed, for instance, that certain birds build nests and sing songs in exactly the same manner as their parents and, moreover, will do so even when they have been reared in total isolation. Patterns of behaviour which appear of their own accord are called **instinctive** or **innate.**

Instinct is a term commonly misused. We say the driver of a car instinctively brakes when a child runs off the side-walk. What we really mean is that he brakes without thinking. By tuition and experience he has learnt to make an emergency stop as soon as danger threatens. If a person were really able to brake 'instinctively' he would be able to hit the brake pedal the very first time he sat in the driving seat, and without taking thought. Instinctive behaviour, by definition, is inherited from previous generations and is rigid, so that it varies little from one individual to another within a species.

Wishful drinking
Many people have the commendable habit of putting out a dish of water for birds to drink. Sometimes they forget to replenish it and one man was rewarded for his forgetfulness by seeing a favourite bird, that habitually came to drink, giving an exhibition of **vacuum activity.**
He saw the bird alight on the rim of the dish and take beakfuls of water, throwing its head up in the usual way to swallow the water. The thought occurred to him that there could be little water in the dish. It was, in fact, bone dry. The bird had been carrying out a series of actions in the absence of appropriate stimuli. Had there been water in the dish it would have drunk some. Its memory of having drunk at that spot stimulated it to go through the drinking action *in vacuo.* German scientists call this *Leerlaufreaktion,* for which 'vacuum activity' is a not very satisfactory translation. As has so often happened, a piece of animal behaviour has been identified and studied first in Germany; then comes the difficulty of translating appropriately.
It might be suggested that the bird had come to associate the man with the provision of water and chose to remind him to refill the dish in this simple way. That would be to commit the heinous sin of being 'anthropomorphic' (p 88). Yet when it comes to interpreting animal behaviour it can be the case that 'your guess is as good as mine'. A dog will sometimes express his desire for food — whether because his meal is late or because he wants to share something somebody is eating — by standing in front of that person, looking up into his face and repeatedly pushing his tongue in and out. The gesture seems quite delib-

A newborn baby instinctively moves its head in search of its mother's nipple. It is born with this ability and all babies behave in the same way. Moreover, it does it in the same way no matter how often it repeats the action. Because instinctive behaviour is basically the same for all members of a species, the study of a wide variety of animals has shown how such behaviour patterns are adapted to a particular way of life.

The second school of behaviour study has its centre in America. Whereas Lorenz and his followers preferred to study wild animals in natural situations, the American school has concentrated on the study of white rats, cats or chicks in unnatural, but very closely regulated, laboratory conditions. Of more importance, they concentrated on learning, that is the ability of an animal to modify its behaviour in the light of experience. Its behaviour undergoes a permanent change as opposed to the temporary changes due to motivation. One of the main exponents of **behaviourism** was J.B. Watson who considered that very little behaviour is inherited. Every animal, and human being, can be considered as being born as a blank notebook on which experiences are then recorded.

Both European instinctive and American learning schools have done much to explain the mechanisms of behaviour, so we shall consider some concepts of instinctive and learned behaviour in turn.

Instinctive behaviour   Instinctive patterns of behaviour are relatively simple and inflexible so they can be 'dissected' into their component parts. Instinctive behaviour has the same simple basic arrangement as reflex behaviour in which a stimulus is sent to the brain where the appropriate response is ordered. Quite how the brain links the stimulus from the sense organs with the response by the effector organs (limbs, jaws, tail etc) is not known. Ethologists have postulated an **innate releasing mechanism** (IRM) in the brain that identifies the stimulus and releases the correct behavioural response, which may then be guided by orientating stimuli. The sudden appearance of a predator frightens a frog into jumping and blue light guides it to the safety of water. An example of instinctive releasing stimuli is a newly hatched chick which has to find its own food and water after hatching. It has a tendency to peck at three-dimensional objects which contrast with the background or at round, bright, reflecting surfaces, and so is guided to seeds and water droplets.

The innate releasing mechanism is more of an abstract concept than an actual mechanism but it is useful to have this

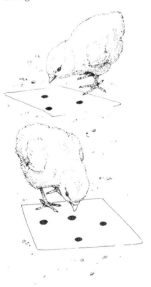

An experiment demonstrating the pecking behaviour of newly hatched chicks: they have an immediate tendency to peck at any round bright dot or small object that contrasts with its background.

concept of a selecting and triggering device for each item of instinctive behaviour. The IRM can be modified by a variety of factors so that the response to a stimulus is flexible.

First, to indulge in any activity, an animal must have the necessary motivation. That is, it must be in the mood to perform this activity. A starving dog gulps its food avidly. If it is given a second helping it will eat it more slowly and may leave a few scraps. The third helping is ignored because the dog has reached satiation. The dog's motivation for eating has changed during the course of its meal. When replete, it lacks motivation, but as time passes the food is digested, the energy is used up and the dog becomes motivated again. Motivation can also be suddenly raised even in a replete dog by offering a special titbit.

Motivation for a specific activity is often called a **drive,** for instance, a hunger or a sex drive. We cannot measure drives directly but we can relate the strength of the feeding drive, for instance, to the time that has elapsed since the last meal or to the amount of food the dog eats before it is sated. It is probable that each drive is related to a physiological state. The hunger drive may depend on the level of sugar in

Two cats meet on a fence at the boundary between their domains: unable to resolve the conflict between aggression and the desire to retreat, one cat resorts to the quite irrelevant behaviour of grooming itself — a displacement activity.

The signal for birds to start their nesting activities in the spring is the gradual lengthening of the days. Somehow the sun affects the birds' sex glands through the eyes and sex hormones are produced. Androgen hormones lead the male to sing and display. In the female, oestrogen secretion started by sunlight is enhanced by sight and sound of the male. As a result eggs start to develop and she begins to collect nest material. As the nest takes shape, it, too, becomes a stimulus for the production of sex hormones. When it is nearly finished, the female becomes receptive for mating; but before she lays her eggs she must finish the nest. In canaries, the species used to unravel this story, the female sheds her breast feathers to form a warm brood patch for incubating the eggs. Oestrogen and other hormones trigger the loss of feathers, and the rubbing of the nest on the bald patch acts as another stimulus, maintaining the production of sex hormones. So there is an interlocking cycle of events. The presence of the male is a continued stimulus, but the brood patch is very sensitive to the feel of the nest bowl. This initially results in the collection of feathers to line the nest just before the eggs are laid; after they have been laid, the contact with the nest maintains the female canary's drive to incubate.

the blood or the emptiness of the stomach, while the sex drive may depend on levels of sex hormone in the blood.

When an animal is activated by a drive, it seeks to achieve the satisfaction of that drive. Three phases of behaviour can be seen in seeking satisfaction. First, the animal demonstrates **appetitive behaviour** and searches for a goal: food, water, warmth or sexual partner. A hungry dog has to look for food and, in the wild, it has to seek, hunt and kill prey. It may be guided by a **search image.** We can find a lost object more easily if we know what it looks like. A bird may start to feed on a particular kind of insect and, having got this image into its head, ignore others which look different even though they may be edible.

Having found the goal, the animal engages in a **consummatory act.** It eats, drinks, mates and so on. Lastly, its goal having been achieved, the animal's motivation is dissipated and the drive is rendered inactive. The animal no longer responds to the stimuli that set its appetitive behaviour in motion.

Sometimes two opposing drives are activated at the same time, leading to a **conflict situation,** as in the case of the jewelfish (Chapter 7), which had to choose between a worm and one of its babies. In this instance, the conflict was satisfactorily resolved but a bird defending its territory boundary may have such a conflict between a drive to attack and a drive to retreat from its opponent that it may be diverted into a completely irrelevant act, a **displacement activity.** It may tug at grass, preen itself or even fall asleep in the middle of the violent dispute.

The role of hormones There is a link between motivation and hormones but the exact role they play is difficult to assess. Outside the breeding season the sexual behaviour of male songbirds dies down and finally disappears as their testes shrink. In spring, the testes increase in size, they once more secrete the sex hormone testosterone and the cycle of the breeding season — singing, courtship and nest-building — starts again.

The action of hormones is relatively slow when compared with the speed of nervous responses. For instance, the pigment cells in the skin of a cuttlefish, which are under nervous control, produce a change of colour in a split second, while a chameleon whose pigment cells are controlled by the hormone intermedin, secreted by the pituitary gland and carried around the body in the bloodstream, takes several minutes to change colour.

That hormones play a part in reproductive behaviour has been known ever since domestic animals, and men, have been castrated. The testes produce the hormone testosterone which is essential for the maintenance of the sex drive and a castrated animal will very quickly show sexual behaviour if testosterone is injected into it. Similarly, removal of the ovaries abolishes female sexual and parental behaviour. Hormones are involved in the gross forms of behaviour which we call **emotions.** Aggressive behaviour, rage and fear are influenced by hormones secreted by the reproductive organs and the adrenal cortex. The latter secretes adrenalin and nor-adrenalin, which is chemically slightly different from adrenalin. These hormones prepare the body for fight or flight, that is they increase the heart rate, mobilize sugar in the blood and generally make us alert for sudden activity.

Rhythmic behaviour Many animals perform **rhythmic behaviour.** Diurnal animals are active during the day and sleep during the night whereas nocturnal animals do the reverse. **Crepuscular** animals are active at dawn and dusk. Patterns of behaviour may be shorter, as in the four-hour cycles of some small mammals, or longer, as in the yearly migrations of birds. How the cycles or rhythms are controlled is something of a mystery. One of the basic factors is the animal's **photoperiodism,** its response to light and dark. Over and above this, it is thought that each animal has a **biological clock.** Even when animals are kept in continuous light or darkness it has been found that the clock continues to operate their normal rhythmic behaviour. Moreover, even isolated tissue cells taken from an animal continue to show rhythmic activity and rhythms are often independent of temperature. Cold may slow down an animal's activity but its rhythms of activity are unaltered.

Cockroaches are most active during the early part of the night and, if placed in darkness, they continue to show a peak of activity about every twenty-four hours. In practice, the daily periodicity of constant rhythms of behaviour is usually a little longer or a little shorter than twenty-four hours, and they are therefore known as **circadian cycles** (Latin *circa,* about, *dies,* a day). The cockroach cycle is a little less than twenty-four hours and, if kept in continuous darkness, cockroaches become active a little earlier each day. In natural conditions the cycle of light and darkness resets the internal rhythm and this is necessary to adjust the cockroach's activity to the changing length of night throughout the year.

Signalling sequence
Lizards of the iguana family have a head-bob or push-up display. On meeting another lizard, each bobs up and down, bending and flexing its forelegs and raising and lowering its head like a miniature gymnast. Such fixed action patterns (FAP) are often used as signals between animals. Head-bob displays are used by sparring males in courtship and are given when an unreceptive female rejects a male. Each kind of iguana has its own rhythm, so that an anole can be distinguished from a Desert iguana and a Desert iguana from a chuckwalla. However, within the species, each individual has its own, personal sequence of bobbing. The first four seconds of each head-bob display by a chuckwalla, a lizard of the American deserts, is called the 'signature' and is unique to that individual.

An important feature of the head-bob signature is that although it may stop in the middle, it is always started at the beginning of the sequence. The nervous mechanism controlling the fixed action pattern is so rigid that it can only transmit 'orders' to the muscles in a set series, starting at the beginning.

Strange things happen to FAPs when species interbreed. Lovebirds are small parrots which breed readily in captivity. Some species carry strips of nest material in their bills, but others tuck strips into their rump feathers. Hybrids between the two inherit a mixture of both FAPs for carrying strips. The result is confusion; the birds do not know which way to carry the strips and it takes two years for learning to overcome the inherited mix-up.

Another little understood phenomenon is that usually referred to as a **time sense.** It is doubtless linked with this hypothetical clock but concerns events that are regular but less than daily. A pet animal soon learns if something pleasurable, such as special food, is given it regularly every Tuesday, say. It shows by its actions that it is aware when Tuesday has arrived.

Fixed action pattern The response to a stimulus in a sequence of instinctive behaviour is often in the form of a stereotyped and inherited series of movements, a **fixed action pattern,** an amalgam of reflexes, kineses and taxes. The classic example of a fixed action pattern was investigated by Lorenz and Tinbergen. Birds nesting on the ground, such as geese, usually make shallow nests with low sides. Eggs are very liable to be knocked out of the nest but they are raked in by the sitting bird hooking its bill over the egg and rolling it back. The shape of the egg makes it roll awkwardly and the bird steers it with its bill. The action is stimulated by the sight of an egg outside the nest and is compounded of rolling and steering. In the first part of the action the bird draws the egg towards the nest with its bill. If the egg slips out from under

Sparring male chuckwallas using the head-bob display (see opposite).

27

the bill, the movement is completed in its entirety before a new attempt is made. The side-to-side steering movements of the bill are superimposed on the rolling movement to keep the egg moving in a straight line.

Fixed action patterns are inherited but their performance may improve during the animal's life. **Maturation of behaviour** comes with development. It might be thought that a young bird has to learn to fly but some birds such as Sand martins and puffins are reared in burrows where they cannot exercise their wings. Nevertheless, young martins and puffins fly reasonably competently on the first occasion that they launch themselves into the air. The ability to fly appears when the bird has reached a certain stage in its development and practice is needed only for the fine control of flight. Newly hatched chicks, with their immediate tendency to peck at objects that contrast with the background, improve their aim over the course of a few days even when given no practice. The maturation of this behaviour has been due to development of the nervous system.

Learning

An animal learns, as we have seen, when its behaviour changes as a result of experience. For instance, it learns to avoid harmful situations or to find good food supplies. The chick that has an instinctive tendency to peck at objects with certain physical characters also learns what is edible. It learns to peck seeds but ignore small pebbles. Experience has modified its behaviour to make it more efficient.

Learning of some form or another can be found in almost every corner of the Animal Kingdom, even among simple animals. The most elementary form of learning is **habituation** in which the animal 'gets used to' a repeated stimulus. If a snail is gently tapped on the shell it withdraws its 'horns' or antennae. Repeated tapping results in the horns being withdrawn less and less until there is no response at all. The snail is now habituated to the tapping and 'ignores' all but a severe jolt. Habituation is a common feature of human life, as when we become used to a regular noise. The value of habituation is that it enables an animal to ignore unimportant stimuli. It would be a disadvantage to a snail if it had to withdraw every time it was hit by a drop of rain during a downpour and if soldiers did not get used to gunfire, all battles would end in stalemate. To survive, a soldier has to learn to react only to explosions that could be relevant to his personal situation.

Familiarity breeds safety
A honeybee worker spends the first three weeks of her adult life working inside the hive. She then joins the foragers quartering the neighbourhood in search of the pollen and nectar. However, before she becomes completely a forager, she will have left the hive to make short exploratory flights, during which she learns the position of the hive by relating it to nearby landmarks and the position of the sun. So, by the time she becomes a forager, she will be able to find her way home no matter where her searches take her. This is an example of **latent learning** (p 31). She gains no immediate reward for her actions, but the information is stored for future use.
Through exploration and orientation, animals living in one place get to know its topography intimately. A mouse put into a new cage is at first very restless. It appears to be trying to

escape, but is in fact investigating its new quarters. Every nook and cranny is examined with eyes, ears, nose and whiskers. At first it moves cautiously, then it wanders about with more confidence until finally it scampers from one corner to another without hesitation. Now, a sudden alarm will send the mouse to cover in a trice. It knows its cage so well that its mad rush is not blind, every footstep has been worked out in advance.

A remarkable case of latent learning is shown by a fish *Bathygobius soporator* living in rocky tidal pools. It makes its way unerringly about the shore by jumping or slithering from pool to pool. It cannot see from one pool to the next but it never takes a wrong turning and never gets stranded. It learns the topography of its home area by swimming over it at hightide.

Association   This is a form of learning in which the animal learns to associate a cause and an effect. The classical experiment to demonstrate association consists of shutting a cat or dog in a box. In its efforts to get out, the animal accidentally knocks a lever which releases the door catch and it escapes. When the procedure has been repeated a few times, the cat or dog learns to associate knocking the lever with escape and it can get out as soon as it is shut in.

A form of association is the **conditioned reflex** discovered by the Russian physiologist I.P. Pavlov when he was studying the secretion of digestive juices in dogs. As we all know, the first mouthful of food causes our mouths to water. The saliva that floods the mouth acts as a lubricant when food is chewed and swallowed, so salivation is the first part of the digestive process and is a straightforward reflex consisting of stimulus eliciting a response. Pavlov found that dogs behaved in the same way and that if he rang a bell or flashed a light before presenting the food the dogs soon learned to associate these with eating. As a result, the dogs started to salivate as soon as they heard the bell or saw the light, even if they later received no food. This is the conditioned reflex, the bell and light being the conditioned stimuli.

What appears to have happened in the animal's mind is that the response — salivation — appropriate to one stimulus, food, becomes associated with a new stimulus — bell or light. Technically, the presentation of food is called a **reinforcement.** If the bell is rung too often without any food being given, the dog stops salivating. The conditioned reflex has become attenuated and finally becomes extinct. To maintain a conditioned reflex, the reinforcement has to be repeated at intervals.

Pavlov's dogs were examined in the clinical atmosphere of the physiological laboratory. They were strapped into position (so that the saliva could be collected) in a soundproof room. Their behaviour was, therefore, cramped and unnatural but conditioned reflexes do appear naturally. A dog soon learns to salivate when it sees food and, later, merely on hearing a tin being opened or its food bowl being rattled. These are conditioned stimuli.

A second kind of learning by association is called **instrumental** or **operant conditioning** or, in everyday English, **trial-and-error.** Instrumental conditioning has been studied intensively by the American psychologist B.F. Skinner, who is famous for his invention of the 'Skinner

box'. The box is a soundproof container fitted with a lever and a food tray. A rat is placed in the box and, during its exploration of its new surroundings, it chances to press the lever. Immediately a pellet of food is dropped into the tray. After accidentally pressing it a few times, the rat learns the connection between the lever and the appearance of food. It will now deliberately work the lever to obtain the reward. Skinner boxes present rats with a very unnatural situation but this trial-and-error learning is widespread in nature. Any animal searching for food is learning by trial-and-error. At first, a young badger wanders about investigating everything that comes its way. It may tip over a stone with its snout and find worms and centipedes lurking underneath. It soon learns that turning over stones is rewarding. At the same time it stops searching in places that lead to little or no reward and eventually it becomes a skilled forager.

The reinforcement in association learning can be a reward or a punishment and is the basis of animal training whether of pets or circus animals. A puppy is housetrained by teaching it to associate a pat with 'good' or 'bad' behaviour. However, an animal can learn without any obvious reward or effect on its behaviour. The learning remains dormant in the

The House sparrow is numerous and widespread over most of Europe and a large part of Asia as well as North Africa. It has been introduced into South Africa, Australia and New Zealand and also into North America where it has spread extensively. This distribution indicates two things: that the bird is adaptable and that human reactions to it are not wholly adverse.

The charm of the House sparrow is its adaptability. To give one example, a group of sparrows were seen catching damselflies, relatives of dragonflies which in their early life live under water in ponds. When ready to change into the adult insect, the nymphs of the damselflies climb up stems of the water plants, their outer skin splits and the damselfly comes out. The sparrows had discovered this unusual bounty and spent the day over and around the pond, picking off the damselflies as they emerged. Some of them were not easy to catch. The sparrows had to hover laboriously to locate a damselfly, and then run on a soft raft of blanketweed that sank beneath them so they had only a split second to seize an insect as the water rose up to their knees.

In part, this adaptability of the House sparrow arises from its **imitativeness**. It has been seen trying to hover over water like a kingfisher, or hanging upside down at the birdtable like a tit in order to get food. On one occasion, a wagtail was hopping from one leaf to another of a water lily in a pond, pecking off insects. A little later a hen sparrow was seen attempting to copy the method used by the wagtail.

animal's brain until it is required. This is called **latent learning.** Most animals gradually learn their way around a territory or a home range, finding out where there is food or shelter from enemies. At the time, this information may have no particular use. If, however, a hawk should swoop at, for example, a Reed warbler, a blind rush for cover may result in the warbler becoming caught in a dead end, but an intimate knowledge of its reed bed will allow it to slip through the dense growth without hesitation.

Insight   Latent learning leads to the highest form of learning, involving insight — the appreciation of relations. The classic instance of insight learning is that of Archimedes, the Greek philosopher, who sat in his bath pondering how to ascertain that a crown was of solid gold rather than of silver gilt, without damaging it. The answer suddenly dawned and, as everyone knows, Archimedes leaped out of his bath and ran through the streets naked, shouting 'Eureka!'. The essence of insight is that the answer 'comes in a flash'. There is evidence that animals sometimes indulge in insight learning where they suddenly produce a new response too rapidly to have worked it out by the laborious process of trial-and-error. Indeed, the solution is reached without any physical activity, let alone the making of errors, so that the animal must have been putting together the necessary data in its head.

Sometimes an elephant will pick up a stick in its trunk and scratch itself. It could scratch itself with its trunk but it seems that the elephant has insight into the superior scratching properties of a stick. It is tempting to suppose that such an act shows that the animal is having ideas or is reasoning. This would be a dangerous assumption without knowing more about the mental processes; but even some insects can show insight. If the clay nest of a Mason bee or Potter wasp is artificially damaged, the insect will inspect the damage, fetch some clay and repair it, thus demonstrating an appreciation that all is not as it should be and knowledge of how to set it to rights.

The division of behaviour into instinct and learning or inherited and acquired is convenient but very misleading. There used to be considerable argument, which is not yet dead, as to the roles of instinct and learning in the lives of animals. This is the so-called nature-nurture argument. Is an animal's behavioural capability fixed before birth or is it born with a mind like a blank ingot which is hammered into shape in the forge of experience? Recent researches suggest that this is a non-argument and that an animal's behaviour is a mosaic of learned and unlearned components.

Indeed, it is becoming very difficult to define what is an instinctive piece of behaviour. It might be expected that the behaviour of a newly hatched chick must be instinctive because it has had no experience whatsoever, but the chick was alive while still within the egg. Both its senses and its muscles were working. In some species there is a two-way vocal communication between the unhatched chick and the parent brooding it, so that they learn the sound of each other's voices, and also between the chicks in the clutch, so that they all hatch together. All react to alarm calls by ceasing to call. Meanwhile movements of the bill, neck and limbs are preparing the chick for the time when it has to walk and peck.

Once hatched, the chick's behaviour starts to change through experience as we have seen. Both instinct and learning join to effect its survival. The chick would not live if it did not leave the egg with a built-in knowledge of how to feed itself but it would waste time, and get indigestion, if it could not learn what is edible. The song of the White-crowned sparrow of the Pacific Coast of North America varies from place to place. Each male sparrow is born with a simple song that it produces even when reared alone in a sound-proof box, but it has to learn the additional notes of the local dialect by listening to nearby males singing. It does this when it is less than three months old and stores the memory until it is mature and ready to sing.

Newly hatched gull chicks beg for food by pecking and grasping the bill of the parent who then regurgitates food for them. They hatch with the ability to peck and a notion of the essential features of the parental bill: shape and colour. This enables them to beg for food, but with experience their pecking becomes more accurate and they also learn in more detail what to peck at. Furthermore, before their aim improves, they accidentally strike food held by the parent and so learn what is edible.

The distinction between inherited and acquired behaviour is also blurred by the fact that the ability to learn is inherited. Some animals readily learn a pattern of behaviour which other species can never master however much they are trained. For instance, some birds are quite unaware of the identities of their offspring, even of the number in the brood. Eggs and chicks can be added or removed from the nest without any effect. Guillemots which lay their eggs on

Second childhood

An English zoologist had three pet Harvest mice, each less than 3 inches long including the tail, and weighing $\frac{1}{6}$ oz. He kept them in a large vertical glass container furnished with tall grasses and coarse herbage, a microcosm of their natural habitat. This cage stood on a table in the sitting room, so the zoologist could not help seeing them daily and many times a day. After a while he began to be aware that his pets could be seen actively climbing and feeding only at certain times.

One Sunday it rained heavily all day. The zoologist spent it in an easy chair beside his Harvest mice, with a notebook, taking notes of when they were asleep, when awake and what they did when not sleeping. At the end of sixteen hours he had found the Harvest mice slept for three hours, were active for the next three and so on. He checked on this on other days and also stayed up all night to watch what happened then. It transpired that, at about this same time, other zoologists were finding that House mice, shrews and other small mammals also had an **activity rhythm** of three hours asleep, three hours active.

The Harvest mice in question lived for four years, which is much longer than had been previously known for that species. This was certainly due to the secure conditions they enjoyed in their cage, with ample food, no enemies, and no extremes of heat or cold. During the last few months of their lives their behaviour changed. They seemed to be doing things they had done when first they were put, as juveniles, in the cage. They played with each other more and seemed to be in a second childhood.

crowded cliff ledges without any form of nest learn to recognize both egg and chick. Flamingos and gulls do not learn to recognize their eggs, which cannot get muddled with the neighbours; but do learn to recognize their chicks which run around and mix with others. As a general rule, then, an animal's learning capacity is related to its particular needs.

Memory   Retaining information is an integral part of the learning process. An animal cannot benefit from experience unless it can retain a memory of it. There appear to be three mechanisms involved in memory — recording, storage and recall — but little is known about them. It seems that forgetfulness could be due to a breakdown of recall rather than of storage because **hypnosis** allows recall of 'long-forgotten' events. How memories are stored has been the subject of some speculation and investigation. The current theory is that there are two stages to memorizing. Short term memory is set up by electrical circuits in the brain and can be destroyed by a powerful electric shock. This is translated into long term memory by the manufacture of specific proteins. Electric shock now has no effect on the memory. The distinction between long and short term memory is seen in elderly people who can recall childhood scenes but are vague and forgetful about current events.

Learning from others

So far, behaviour has been considered as being inherited in a rigid fashion from the parents or acquired through personal experience. But some behaviour can be acquired from parents or others by shared experience. At the simplest level, this need not involve actual learning. In **social facilitation** the animal follows the behaviour of another only as a temporary measure. Thus, if a chicken which has fed to satiation is placed with a hungry chicken, and both are given food, it will start to feed, even though it is not hungry. But if the hungry chicken is removed, the sated one stops feeding.

A more permanent change of behaviour occurs in **observational learning** where one animal continues the new activity after its 'teacher' has been removed. Observational learning by kittens has been demonstrated by setting three groups a task (for example, pressing a lever 20 seconds after a light has flashed). Kittens that had no experience of this task failed to learn how to do it but those who had watched adult cats performing it learned after several days' practice. The most interesting point to emerge was that kittens who watched their own mothers learned four times as fast as

33

those who could only watch strange females. From these and other experiments it is clear that the mother-child bond is very important in the education of young animals. Indeed, young animals usually learn by following the lead of their parents, rather than by being actively taught.

Observational learning is probably especially important among wild animals living in groups. Tits moving in a flock through a wood learn from each other the best places to look for food and young Wood pigeons feeding in a flock on a field watch their elders to see what is the best kind of food.

Although observational learning is a good behavioural 'shortcut', nearly all animals learn no more and no less than previous generations of their species. They belong to stagnant societies like some human societies in which there have been no new ideas, so that art, manufacture and social life reached a peak and never advanced beyond it. Development of new ideas and new techniques is **cultural advancement,** often held to be a purely human character. But animal societies, too, show signs of a cultural life, in the transmission of behaviour from one generation to another by learning. Japanese zoologists making long term studies of wild Macaque monkeys have found several instances of

Macaque monkeys in Japan discovered for themselves that washed sweet potatoes are more palatable than those that are sullied with earth.

As the days of the last quarter of the October and November moons dawn over the South Seas, boats gather offshore to collect the annual crop of the Palolo worm *Eunice viridis*. This worm, a relative of the ragworms, lives in burrows in coral reefs; but every year the rear part of the body alters drastically. Reproductive organs develop and the limbs become paddle-like. Finally, this part breaks free from the rest of the animal, swims to the surface and ruptures, releasing eggs and sperms. The incredible part of this strange story is that all Palolo worms in one area break in two at exactly the same time. The masses of eggs lie so thickly at the surface that they can be netted by the South Sea islanders as a delicacy.

Spawning is limited to four days, the last quarter and preceding days of the October and November moons. The related *Eunice fucata* of the West Indies spawns in the third quarter of the June and July moons and another worm *Ceratocephale osawei bochi* of Japan spawns at the new and full moons of October and November. When an animal releases its eggs and sperms into the sea the chances of fertilization are increased if all individuals do so at once. Synchronized spawning is therefore common among marine animals; but few spawn within such a narrowly predictable range as the Palolo worm. On Savii island the spawning is predicted, three days beforehand, by the mass migration of land crabs to the sea, also to spawn. Controlling the yearly rhythm of the Palolo worm is a **biological clock** (p 26), which is set by the cycles of the moon and seasonal changes in light intensity, and triggered by hormones.

cultural transmission. To make the monkeys easier to study, food in the form of sweet potatoes and grain was provided regularly. Certain individuals discovered that sweet potatoes tasted better after being washed in the sea and that grain could be separated from the sand of the beach by throwing it into the sea where the sand sank and the grain floated. These new 'ideas' gradually spread through the rest of the troop.

The evolution of behaviour

Living organisms probably began in a very modest way when, in a world of lifeless rocks, certain chemical elements came together to form complex molecules. In due course collections of these came to be imbued with the ability to feed and reproduce. First came plants able to manufacture their own food with the aid of sunlight. Then came animals that fed on the plants. Both were simple and microscopic in size. Relieved of the necessity of being each their own food factory, the animals developed locomotion to find their food and a sensitivity to detect and select it. All this was at a low level of structural organization and behaviour.

Today we see in the vast panorama of animal life, progress from microscopic forms to the highly specialized birds and mammals, largely through the development of a nervous system and of part of this system to act as a control centre, the brain, with an almost infinite variety of behaviour patterns, at every mental level.

These behaviour patterns have evolved by the processes of natural selection in the same way as the physical characters. Both are said to be adaptive, that is they suit the animal to a particular way of life. Sharp claws and teeth and supreme agility fit a cat to a flesh-eating way of life. So, too, does its hunting behaviour: the recognition of suitable prey, the stealthy approach, the final pounce and bite. Unfortunately for the study of their evolution, behaviour patterns cannot be fossilized like bones. However, we can study behavioural evolution by examining the differences and similarities between animals now living.

The limb bearing five digits shows the common ancestry of amphibians, reptiles, birds and mammals, even though the original plan has been modified to form wings, flippers, hoofs and so on by a variety of evolutionary pressures. The basic plan has withstood these changes and, in the same way, so has the behaviour of reptiles, birds and mammals when scratching their heads. Nearly all species scratch with the hindlimb brought over the shoulder and it is impossible to train them to do otherwise. Conversely, some traits are readily modified to suit changing conditions.

Many ground-nesting birds rely on camouflage to protect their eggs from predators. Not only do the eggs merge with the background but the adults take active steps to conceal the nest. Droppings and empty eggshells are taken away, not out of cleanliness but because if left they would betray the nest. Both these habits are common in the gull family whose nests are robbed by foxes, crows and other gulls. There is, therefore, an evolutionary pressure to camouflage the nest: those gulls that neglect this task fail to raise offspring. But some gulls have taken to nesting on cliff ledges where many predators cannot reach them. One of these, the kittiwake, differs in some thirty-three behavioural and anatomical features from ground-nesting gulls. In many ways it is still very clearly a gull in behaviour and appearance but, having nothing to fear from foxes and other earthbound predators, the kittiwake has relaxed its camouflage. Droppings and eggshells are not removed. But cliff nesting now poses new problems. Kittiwake chicks have to avoid falling off the ledge, so, unlike the chicks of most gulls they stay in the nest until ready to fly. The nest itself is a deep solid cup and the young kittiwakes have sharp claws for holding on.

Intelligence

It would be useful to be able to measure quantitatively the intelligence of animals. Intelligence is essentially the ability to learn but animals cannot be compared by means of intelligence tests because their brains work in different ways. There are similar problems even in studying human intelligence. A difficulty in setting intelligence quotient or I.Q. tests is that they must be equally applicable to all people and not favour those who can read well or are trained in a particular skill.

A rough idea of intelligence can be gained by looking at brain size or, more precisely, at the size of an animal's brain compared with its body size. Among arthropods, it is the insects that have the largest brains for their body size and, among molluscs, it is the octopus, squid and cuttlefish. This accords with our knowledge of the mental capacities in these two big groups. Among the vertebrates there is a gradation in brain size and structural complexity from the lowly fishes through to man. Anatomical comparisons are, however, not always reliable. Flies and bees have brains of similar size but bees are much more adept at learning.

Rope tricks
Many people have observed small birds hauling in a string threaded with peanuts that has been hung on the birdtable or on the branch of a tree. There are also a number of records over the past few years of members of the crow family hauling bones up to a perch, by pulling the string on which the bones were suspended with the beak, holding the loop with the foot, taking another loop with the beak, and so on.
These actions have been generally regarded as the result of **insight learning**. Proof that they are not merely instinctive actions came in 1957 when, in both Norway and Sweden, Hooded crows were found to be stealing bait and fish from lines set through holes in the ice. A crow would take a line in its beak and walk backwards from the hole. Then it would walk forwards again, carefully treading on the line to prevent it slipping back. It would repeat this until the fish or bait was drawn to the edge of the ice, when it would seize it.
This elaborate set of actions to achieve an end is

more worthy of being called **problem solving**: a more recent example took place when a householder hung a piece of fat on a string from the bough of a tree. The fat was intended for the Blue tits and Great tits, gymnastic birds that can hang upside down and peck at a piece of fat. A gull arrived on the scene, walked over and stood looking up at the fat. It then made a flapping jump and by dint of effort was able to take small fragments of the fat. A crow arrived and perched on the branch. It took in the problem more or less at a glance, hauled the string up with its foot, using the tactics described for the Hooded crow, and very soon was able to seize the whole of the fat.

Among mammals, brain/body ratios give quite the wrong impression. Shrews have brains proportionally twice as large as those of elephants, but no one suggests that shrews are twice as intelligent. If anything, it is the reverse, and the answer lies in another important feature of the brain: whether the surface is convoluted or thrown into folds to increase its surface area. The surface of the brain of a shrew is smooth, that of the elephant is heavily convoluted.

Nobody had ever dissected the brain of a porpoise until R. H. Burne did so at the Natural History Museum in London, in 1935. The zoologists who saw the dissection all expressed surprise, saying that from the convoluted surface you would expect porpoises to be highly intelligent. A few years later the first seaquarium was built in Florida, for keeping porpoises in captivity. Subsequent study of these animals showed that whales, dolphins and porpoises could be next to man in intelligence, but we must be very careful not to fall into the 'Clever Hans' error. An experiment involving two dolphins cooperating to carry out a task while in contact only by sound seemed to suggest that they must have communicated with each other; but each could have known what the other was doing in the same way as we know what someone is doing in the next room by listening to the sounds of activity. On the other hand, orang-utans were thought not to be so intelligent as chimpanzees because they did not do well in certain tests. Then it was realised that the orang-utans were handicapped because their hands were not so dextrous as those of the chimpanzees!

'Intelligence' is a word commonly used and generally understood, although a precise definition is virtually impossible. One definition that commends itself is that intelligence is the capacity to recognize a problem, to find a solution and to act quickly to put that solution into effect. The third part of the definition is perhaps the most important; otherwise it is difficult to draw the line between simple problem solving and what we normally mean by intelligence. Insight behaviour (see p 31) must also be a closely related process, and all three processes need the ingredients of reasoning and thinking, to which the process of learning must make a considerable contribution.

In this age of computers it is fashionable to compare the working of these remarkable instruments with the human brain, and justly so. Information is fed into a computer and the answer is given out; questions are fed into it and information emerges. Basically this is true for the human brain also, so justifying the view of some people that when we think we are not doing anything more mysterious than calling on our internal computer to work. If we have a knotty problem we can make the effort to find a solution and fail. We can take the problem to bed and 'sleep on it', and the chances are that the answer will occur to us on waking. Our subconscious mind (=computer) has been at work while we slept.

This may be a crude over-simplification, yet it contains enough of the truth to help in our discussion. Just as there are small computers and large computers, so there are small brains and large, and more and less highly organized brains. The size and organization of the brain relate to the rational position of the species in the animal scale. There is, therefore, no *a priori* reason why we should assert, as some zoologists have been known to do, that man thinks, and that no other animal does; or that there is no such thing as intelligence outside the human species. The differences are quantitative and qualitative, and even at the lower levels of the animal kingdom, among insects, for example, there is sometimes the appearance of intelligence, or even of rational thought. Because it seemed absurd to credit insects with such qualities there was a time when such manifestations of seeming intelligence were classified as **plastic behaviour,** so dodging the issue.

On pp 29-30 are described experiments with the Skinner box. A cat put in a crude form of such a box thrashes about and sooner or later one of its paws depresses a lever that allows the lid to fly up. By random action, purely by chance, the cat has found the answer to its distressing problem. The next time it is put in the box it finds the lever quickly. It has profited from experience.

There is a crude comparison to be made between this and our thinking. When tackling a difficult problem we cast about this way and that until, metaphorically, we touch the appropriate lever and the mental lid flies open. The next time we have a comparable problem we solve it in less time. Like the cat we learn from experience.

Fishes' time sense
A remarkable fish story came from New Zealand seventeen years ago. Mr. Wally Ker owned a beach-front property at Double Cove, at the northern end of South Island. On a small pebble beach he started to break up mussels and throw them into the sea. At first a stray snapper came and grabbed a piece and swam away. (Snappers are deep-bodied perch-like fishes living near the shore in warm seas.) Soon, as Mr. Ker continued to distribute his largesse, more and more snappers joined the feast as groups of a dozen or more ventured to within a foot of the shore, their backs flashing iridescence as they broke surface. Before long one in particular would take food from his hand and others became tame enough to allow their backs to be stroked, or would even suffer themselves to be lifted out of the water.

It was mildly interesting to see how the news of a food hand-out spread through the ranks of the snappers; but the most outstanding fact was that in a very short time the fishes had apparently learned that food was being given away each day, at the same spot, and at the same time.

Chapter Two

Senses and Effectors

role of senses: studying the senses — touch: vibration: lateral line —
hearing: ultrasonics: echo-location — vision: insect vision: colour
vision — smell and taste: pheromones: scent marking — sense of
temperature: sense of electric charge: sense of Earth's magnetism:
balance: orientation — how the sense organs work — sign stimulus:
releaser — role of effectors (limbs): intention movements:
ritualization: displacement activity: redirected response

In the previous chapter, the mechanisms of animal behaviour were described. Whatever mechanism is involved, the animal bases its reactions on what is happening around it. Stimulation is the starting point of its behaviour, whether it is a simple trigger stimulus to release stereotyped behaviour or whether it is one that provides the information needed to alter behaviour, that is, to learn. Stimulation reaches the central nervous system (the brain and the spinal cord) via the sense organs and each animal is equipped with the sensory capacity necessary to gather such information as it needs to arrange its life. Understanding these sense organs is therefore essential to understanding behaviour.

The first step in that understanding must be the realization that no two animals have the same sense organs and that our own view of the world through our senses is very different from that of other animals. Disastrous mistakes can be made by assuming that all sense organs work in the same way as our own; a dog, a frog or a honeybee does not see, hear or feel in the same way as we do. Indeed, it requires no particular knowledge of biology to realize this. Anyone can observe that a dog can hear sounds inaudible to the human ear or that a cat can see in 'pitch darkness'.

The lateral line
Fishes, and the tailed amphibians, have a well-developed sense organ of touch or, perhaps more properly, distant touch, which is sensitive to low-frequency vibrations in the water. Called the **lateral line**, the organ consists of a row of small pores running down each side of the body and branching over the head. The pores lead into a canal that has sense cells embedded in its floor. When there is a movement of water around the fish, the fluid in the canals also moves and the sense cells are deflected, triggering nerve impulses to the brain. The lateral line system is used to detect the presence of other animals by the vibrations they set up with their movements; it also detects stationary obstacles because water flowing round them sets up eddies. The position of the source is detected too, because

vibrations reaching one end of the lateral line will be stronger than at the other. The lateral line is particularly important in fishes that live in dark water. Some deep-sea fishes have the organs on stalks to give a greater sensitivity. The lateral line is also used in communication: shoals of fishes keep in formation by its use, and male fishes direct strong jets of water at each other by thrashing their tails, as a means of fighting. Newts use the lateral line in courtship. The male faces the female, bends his tail round and ripples it vigorously to stimulate her lateral line.

Unlike the human eye, an insect's eye is sensitive to ultraviolet light and in visiting flowers insects are sometimes guided to nectar by lines on the petals which show up strongly in ultraviolet light.

Research into the senses of animals has shown that previously inexplicable behaviour can often be quite simply explained. At one time it was baffling how salmon find their way upriver to their birthplace. Now we know it is by a sense of smell. For centuries scholars pondered how bats could catch insects in the dark. Now everybody knows they do so using echo-location. There used to be much talk about a mysterious 'sixth sense' to explain seemingly miraculous behaviour. In the main, animals are using the same senses as we possess but often these are more highly developed. In addition, there are senses used by animals which are outside our experience. Thus, certain fishes communicate by means of weak electric currents and some birds respond to the Earth's magnetic field.

We live in a world dominated by sight. We have good colour vision yet we find it difficult to appreciate that honeybees, for instance, can see ultraviolet light. We may not be able to imagine how the world would look through a bee's eyes but, because we have learned they can see colours outside our spectrum, we can begin to understand how bees locate the correct flowers they need and then unerringly make their way back in a direct line to the hive, often over a long distance.

Sense organs are basically converters of energy. They pick up energy in the form of light waves, sound waves, and the like and they convert it into nerve impulses. This is known as **transduction** and is the same process as when a microphone converts sound waves into an electric current. There are two kinds of sense receptors. The first includes eyes and ears which record information coming from a distance; the second includes the organs of taste and touch, which receive stimuli from objects in contact with the body. A sense organ is made up of sense cells that convert the incoming energy into nerve impulses; it usually has one or more accessory structures that filter or modify the incoming energy. The lens and iris of the eye are accessory structures that respectively focus and adjust the level of light entering the eye.

Studying the senses of animals
It is impossible to ask an animal what it can see, hear or feel so the investigation of sense organs must proceed by a roundabout route, making use of the process known as **conditioning** (see p 29). If, for instance, the aim is to find the range of a dog's hearing, the dog is first trained to come to food at a whistle. Then it is presented with whistles of

increasingly higher pitch and there comes a time when it fails to respond, which means that that particular whistle is just beyond the dog's range of hearing. Because the conditioned stimulus is quite irrelevant to the unconditioned situation, any sense can be tested with a simple response, such as coming to food. The drawback is that the procedure is slow and tedious, particularly for lowly animals which are slow to respond to training.

A more direct way of studying the response of sense organs is to record the impulses running up the nerves from the sense cells. This has only been possible in comparatively recent times. Techniques have advanced sufficiently for the impulses not only in a nerve but in a single nerve fibre to be recorded and so the response of a single sense cell to various stimuli can be measured. The difficulty is that it is necessary to expose the nerves by surgical operation. Nevertheless, with patience, the minutest details of a sense organ's sensitivity can be deciphered and, as the nerves are being tapped, the information so gained is expressed exactly in the form in which the brain receives it.

A blowfly has sense cells of taste on its feet, necessary because a blowfly walks over its food. When these cells come into contact with sugar, say, the blowfly's proboscis is automatically extended ready to suck up the food. Tapping the nerves has shown that the sense organs always react in the same way. There is an initial burst of nerve impulses which gradually die down as the sense organ adapts. The rate of impulses is proportional to the concentration of sugar so that the blowfly can assess its food supply quantitatively as well as qualitatively. Whether and how much it responds to the nerve messages then depends on its own motivation — whether it is hungry or not.

The senses and their evolution

It seems probable that the first sense to have evolved was the sense of touch, although the sense of smell also has a good claim to this honour. We can picture the first living animals doing no more than absorb food from the surrounding water and reproduce by dividing in two. Almost certainly they must have maintained contact with their surroundings by a primitive sense of touch, through a generalized sensitivity of their outer surface. Even now the sense of touch in all animals is still very simple. It is not concentrated in such elaborate sense organs as those of hearing and vision, although since touch can be defined as the detection of slight

Insect's spectrum
The famous Austrian zoologist, Karl von Frisch, demonstrated that honeybees could see colours by training them to feed at bowls of sugar water placed on coloured papers. First he trained them to land on blue paper and they would ignore other colours. Then von Frisch trained bees to respond to other colours in turn until he had found which colours they could distinguish. He also put down papers in different shades of grey to prove that the bees were reacting to colour and not to shade. Bees can see ultra-violet, bluish-green, violet, 'bees' purple, yellow and blue, so their spectrum is different from ours. They can see farther into the blue end of the spectrum but cannot see red. Bees are attracted to flowers by the contrast they make with the background foliage. This is why insect-pollinated flowers are brightly coloured. Although a bee cannot see the red of a poppy, the poppy reflects ultra-violet light strongly. We cannot see this, nor the ultra-violet rim of a daisy.

Nodding birds
Birds have keen eyesight and they rely on their eyes for finding food and detecting danger. Detecting a moving object, whether it be prey or predator, is best achieved if the head is held still, and birds have a remarkable ability to fix the head relative to their surroundings when the body is moving. A kestrel, known as the sparrowhawk in America, hunts by hovering steadily over open ground. It moves slightly as it is buffeted by the wind; but analysis of films has shown that the head is held as steady as can be while the body moves about it. In this way it can detect a beetle

The most conspicuous feature of an otter's face is the set of stout whiskers that serve as organs of touch; they are especially useful in murky waters, when they are extended fully forward, as shown here.

moving through the grass many feet below. Films taken of birds perched in trees that are swaying in the wind show the head stationary in space while the body moves with the swaying of the branch.
Ground-dwelling birds frequently nod their heads as they walk. We are all familiar with the nodding of a pigeon on a pavement, of a coot crossing a lake or of a stately peacock but, in fact, this is an illusion. The bird thrusts its head forwards, then holds it steady while the body catches up. Small, hopping birds do not need to 'nod'. Their heads are stationary between jumps.

pressure, hearing can be thought of as a form of touch. It is the detection of slight pressure waves in the air.

Touch Touch is registered by sense cells or small receptors scattered generally over the body. Relative sensitivity of touch depends on the density of the receptors. Thus, there are more touch receptors on the tips of our fingers than on the backs of our hands and they are at their densest on the tongue, where they are crowded together, which is why a mouth ulcer always feels so large. Sensitivity is increased in many animals by the presence of bristles or whiskers which act as extensions of the body. Whiskers are particularly well-developed in animals that are active at night, like cats, or in those that swim in murky waters, like otters.

Hearing The simplest hearing organs, the sense organs for detecting sound, are little more than touch receptors which are armed with fine hairs for the detection of sound waves. The development of ears in vertebrates and in insects such as crickets has reached a point which allows discrimination of pitch, loudness and direction. Concurrently with this, organs of sound production have been evolved so that animals can communicate over distances. Birds, crickets, frogs, fishes, whales and monkeys are among the many animals which have recognition sounds or songs whose specific patterns act as beacons to attract mates or warn off rivals. Some birds and mammals have a whole repertoire of sounds, each with a different meaning, but man is the only animal with a language proper, with grammar as well as vocabulary.

A specialized development and co-ordination of voice and hearing is seen in **echo-location**, in which the positions of obstacles and of prey are determined by the emission of a series of sharp sounds, the animal then listening for the rebounding echoes. Echo-location, or **sonar**, as it is preferably called, reaches peaks of perfection in bats and dolphins but is also used by seals, the Cave swiftlets of south-east Asia and the oilbirds of South America. A simple form of echo-location is used by the Whirligig beetle. The waves made by its own swimming movements bounce off the bank of the pond and these echo-ripples are detected by its antennae, which are primarily organs of touch. There is reason to believe that echo-location may also be used by penguins, rats and shrews.

Vision We have a biased attitude towards the sense of vision because it is our most important sense and even though we communicate by speech we like also to be able to look at the speaker. As a result, when we try to imagine something, we talk of visualizing it. Because of our reliance on eyesight we sometimes assume that all animals see the same world as we do but, just as we are excluded from the bat's world of echoes, so many animals are excluded from our world of colours, while others can see only a few colours. Giraffes, for instance, confuse green, orange and yellow but can distinguish other colours. Three-dimensional or stereoscopic vision, used for accurate and rapid judgment of distance, is found especially in monkeys (which must be able to grasp branches as they swing rapidly through the trees) as well as in falcons (which need split-second judgment when **stooping** on prey at over 100 mph).

In general the eyes of vertebrate animals fall into two categories. There are those that can see colours and those that can see well in dim light. The former have the retina packed with sense cells called cones, from their shape. Cones are used in colour vision and are also very sensitive to detail. Animals hunting in dim light have eyes containing many rods, sense cells that register light of low intensity but cannot resolve detail. In practice, many animals have both rods and cones, as we do. Nocturnal animals, crepuscular animals and animals that live in the dim regions of the sea are able to let the maximum amount of light into their eyes by having very large pupils. During the day, the pupils contract to shut out excessive light.

Insects' eyes are constructed on a different plan from those of vertebrates and from those of cephalopod molluscs (squid and cuttlefish). An insect eye is made up of a number of units called ommatidia, each having a lens with several sense cells behind it. At one time it was thought that each ommatidium registered on or off, depending on whether or not light fell on it, and that an insect's eye-view consisted of an array of dots, like those in a newspaper photograph. It is now known that the eye is more complex than this and that each ommatidium can register a fair amount of detail. There is, however, a relation between the number of ommatidia and the insect's power of vision. Worker ants that live underground may have no more than six ommatidia in each eye and can do little more than tell light from dark. At the other end of the scale, some dragonflies have twenty-eight thousand ommatidia per eye. They catch their prey on the wing and can see moving objects forty feet away.

Smell and taste The senses of smell and taste are known as the chemical senses. The distinction between the two is

Snake legends
It is a long-standing belief that snakes can hypnotize their victims into immobility. This belief is almost certainly founded on someone seeing a small mammal, such as a rat, stopped dead in its tracks, unable to move, and nearby a snake with its head up as if gazing on it. There is little doubt what has happened: the snake has struck its prey, injecting poison, and the mammal has sought to escape but is paralyzed by the venom.
Even if such a victim escapes into cover the snake can track it, using its flickering forked tongue and its Jacobson's organ. This accessory organ of smell, lodged in the roof of the mouth, and opening into it, is found especially in lizards and snakes. The flickering tongue is pushed out and picks up airborne molecules of scent. When the tongue is withdrawn into the mouth its tips are placed against the opening of Jacobson's organ.
In less enlightened days it was widely but erroneously thought that the forked tongue of a snake was a menace, a venomous organ in itself. It is no more than a highly efficient tool for following trails of potential prey or for seeking out other snakes for mating.

Hermit crabs' long distance detection
On the island of New Ireland that forms part of the Bismarck Archipelago in the southwest Pacific lived a man whose hobby was collecting seashells. In the garden surrounding his house

44

Hermit crabs would wander up from the beach. One evening the collector had placed four shells on a concrete patio. Next morning, one of the shells was missing and in its place was a worn shell. The missing shell was later found under a tree on the lawn with a Hermit crab in it.

Soon afterwards another batch of shells was put on a work bench on the verandah at the back of the house. Two of these precious shells were missing the next morning and worn shells in their place. Again the missing shells were found, each with a Hermit crab in it.

Normally, the Hermit crabs would never come onto the patio or the verandah. For one thing it meant quite a climb up a flight of four concrete steps, and then a journey from the patio to the verandah negotiating a gap of 3-4 inches (the Hermit crab having legs only 1½ inches long), then up the verandah bearer, past the overhang of the floorboards onto the verandah, then up the leg of the bench to get to the shells.

The only way the Hermit crabs could have detected the presence of the shells was by their sense of smell. Yet the shells had been cleaned and only a faint aroma of dead mollusc still clung to them.

somewhat arbitrary but taste usually refers to the sensation caused by chemicals dissolved in water and smell to the sensation from airborne chemicals. Strictly speaking, most of what we call taste is, in fact, smell. The taste of food is really its aroma, which is why food appears tasteless when we suffer from a headcold. Taste is restricted to four sensations: sweet, sour, bitter and salt, each of which is located on a particular part of the tongue.

Until recently, smell was very much of a mystery sense, mainly because we ourselves make so little use of it; but some animals live in a world of odours. They use smell not only for finding food and providing warning of approaching predators but also as a language. Such animals secrete **pheromones**, odorous substances which act like hormones, but outside the body, and are used by animals to communicate with their fellows. The most familiar use of pheromones is when a dog marks a lamp post or tree. In its urine are specific chemicals that make a 'signature' recognizable by other dogs, which can tell the identity and social status of the dog that deposited it. With a bitch, which scent-marks in a less obvious fashion, the pheromones tell whether or not it is in season. Foxes and wolves use a similar method. Many mammals have special scent-producing glands for this purpose. Outstanding are the African and Indian civets which produce a scent from glands under the tail. For centuries the secretion has been collected and used, under the name of 'civet', as a base for manufacturing perfumes. Similarly, 'musk' is collected from the Musk deer of eastern Asia.

Some pheromones do more than act as 'words' in a language of odours. They actually affect the physiology of the receiving animal and are then called **primer pheromones**. If female mice are kept together, their sexual cycles cease; but when a male is introduced they come on heat simultaneously three or four days later. If, then, just after mating, the male is replaced by a strange male, the females lose their embryos and come on heat again. Moreover, the physical presence of a male is not needed to effect this; the same effect is achieved by putting the females into a cage tainted with the odour of a male.

The life of a bee hive is regulated by **queen substance**, a pheromone produced by the queen bee and transferred to the workers when they lick the queen's body. It prevents the workers from becoming sexually mature and stops them making the special brood-cells in which new queens develop,

45

a novel way of ensuring an absence of rivals. Queen substance also attracts drones on the nuptial flight and, when the bees swarm, it is the pheromone that keeps them together.

Temperature sensation   Touch, taste, sight, smell and hearing are our five primary senses but we have other, lesser senses that are important to other animals and there are animal senses which we lack completely. One of our lesser senses is that of temperature sensation. It is an odd sense in that we do not make absolute measurement of temperature as we do with loudness of sounds and intensity of light. When we come in from wintry weather outside, the house feels warm but to someone stepping out of a hot bath the same atmosphere will feel cold. The measurement of temperature is probably relative because the main function of the sense is to bring animals into the most equable conditions, neither too cold nor too hot.

Some animals, however, are guided by heat. Pestiferous animals that suck our blood are guided by our body's warmth. Mosquitoes, fleas and bugs are examples among

Built-in thermometer
A bird's egg has to be maintained at a high, even temperature if it is to develop properly. This is usually carried out by the parent resting the egg on its feet and covering it with its body. The Emperor penguin can maintain its eggs at a level of 34°C even when the air temperature drops to −60°C. The incubator birds or megapodes, however, have abandoned the simple process of using body warmth as a means of incubating their eggs. Instead, they employ heat from the sun's rays, rotting vegetation or even volcanic steam. The best known of the incubator birds is the Mallee fowl which lives in the dry mallee scrub region of Australia, and incubates its eggs in a huge 'compost heap'.
At the beginning of winter,

46

the male Mallee fowl digs a pit fourteen feet across and over three feet deep and fills it with dead leaves. After a soaking of rain, the pit is covered with sand and the leaves start to rot and warm up. Eggs are laid in a chamber in the middle of the rotting heap over a period of four months and hatch at intervals. Altogether, the nest is in operation for eleven months.

During this time the Mallee fowl continually monitors the temperature of the nest. It samples sand from the nest chamber with temperature receptors in its mouth and if the temperature deviates from 33°C, it acts accordingly. In spring, the compost gets very hot, so the Mallee fowl opens the nest to let heat out, but on summer days sand is heaped over the nest to keep off the powerful rays of the sun. On summer evenings and in autumn the nest is opened for the weak sun to penetrate to the eggs.

Rattlesnakes and other Pit vipers can detect very slight changes of temperature in the air through a pit on each side of the head between nostril and eye, which acts as a heat detector.

many known to be guided to their target by heat. Chicken mites feed only at an optimum temperature. During the day, a chicken's body is too warm and the mites lie dormant but its temperature drops a little at night and the mites then come out to feed.

The real specialists in the appreciation of temperature sensation are the Pit vipers, which include rattlesnakes and mocassins. Pit vipers are named for the two pits, each six millimetres deep and three millimetres across, that lie between the nostrils and the eyes. Near the bottom of each pit is stretched a membrane that bears between five hundred and fifteen hundred temperature receptors per square millimetre. It is, therefore, the temperature equivalent of the light-sensitive retina of the eye. The overhanging rim of the pit casts a 'heat shadow' on the membrane and as the 'fields of view' of each pit overlap, there is a stereoscopic effect so that the snakes can detect the range and position of warm objects. The receptors are sensitive to temperature changes of as little as 0.002 degrees C and can detect the warmth of the human hand from a distance of one foot. Thus Pit vipers can pursue and strike accurately at their prey in pitch darkness.

Sensation of electric charges   Every contraction of a muscle is triggered by a weak electric charge that spreads over the surface of the muscle fibres. In certain fishes there are muscles which have lost the power to contract but which can generate a larger electric charge than usual. These muscles are called **electrogenic organs** and they set up a magnetic field, like that around a bar magnet. The fishes also have sense organs that can perceive electric charges, so they can detect any distortion in their own electric field, such as would be caused by a nearby object in the water. The sense organs used for detecting electric fields are minute jelly-filled pits in the skin with sense cells at the bottom of each. Knifefishes of Africa, Asia and South America and the African elephant-snout fishes generate only weak fields but they use them for locating their prey and for communicating with their fellows. Each muscle, or electroplate, as it is called, generates only a fraction of a volt but batteries of thousands of electroplates can generate a powerful charge. The South American electric eel has up to ten thousand electroplates, generates up to five hundred and fifty volts and can stun a horse. The African electric catfish, known to Arabs as *ra'ad* — the shaker, and the Electric ray or torpedo of the Mediterranean can also deliver strong shocks.

Sensation of the Earth's magnetism Some kinds of animals are known to be able to detect the Earth's magnetic field. Such diverse animals as protozoans, flatworms and snails can be led off course by placing a magnet near their line of travel. However, it seems that birds, too, can be affected by the Earth's magnetism. If birds are confined in a cage when the time comes for their migration, they flutter repeatedly against the wall of the cage that faces the direction of their migration route. European robins will continue to do this even in the dark but, if placed in an iron box, their flutterings lose orientation. When the door of the box is opened, they once again take up position as if determined to fly on their traditional migration route. The iron box greatly reduces the effect of the Earth's magnetic field. If birds do have a sort of magnetic compass in their heads, it could explain how they can keep on course even when clouds obscure the sun and stars by which they normally navigate.

Sense of gravity — balance and orientation A sense that is rather taken for granted is that which tells us which way up we are and helps us keep our balance. A sense of balance is not peculiar to man but is present in all animals that move about. In vertebrates, with their well-developed inner ears, it is similar to our own. In lower animals it is simpler. In a prawn, for example, there is a small pit near the base of each antenna lined with minute sensitive hairs. A few grains of sand in each pit impinge on these hairs and indicate to the prawn by their movements when the body is off balance. Prawns shed their cuticle periodically to grow in size, and with the cuticle go the sand grains. A prawn must therefore pick up a few grains, when the moult or **ecdysis** is complete, to drop into the pits in the new cuticle. If a prawn is given iron filings instead of sand, it can be made to turn upside down by holding a magnet over it. We have similar organs of balance. Within the **inner ear**, there are two fluid filled cavities, the utricle and the saccule. Each contains chalky granules called otoliths which lie in a mat of sense cells, the macula. We can stand upright because the two maculae in each ear are continually sending messages, via the brain, to the muscles of the legs and trunk, which automatically correct any movement out of the true. Balance is also maintained with the aid of the three semicircular canals which lead from the utricle. These are fluid-filled and when the head is moved, the fluid moves in the canals and the sense cells lining them record the displacement. If we spin

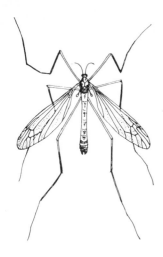

The long-legged cranefly has the second pair of wings reduced to small knobs, known as halteres, that act like a gyroscope.

round quickly, then stop, the fluid continues to race around the canals and we feel giddy.

The gravity sense is useful in keeping balance because gravity exerts a uniform and continuous force in one direction. It provides a good point of reference. Light can also perform this function. Despite our inner ears, we also use our vision to keep balance. It is difficult to balance on one leg with the eyes shut and travel sickness is partly due to seeing our surroundings in motion. For aquatic animals light is a good point of reference because it always comes from above. Some fishes and crustaceans turn upside-down if lit from below.

Two other kinds of **orientating organs** are found in insects. Water bugs use the buoyancy of an air bubble as a point of reference. Water boatmen and backswimmers, for example, are informed which way up they are by the position of a small bubble pressing against their antennae. A sophisticated variation of this is used by the Water scorpion, another water bug. It has two rows of pits in its underside. Each pit is covered by a membrane which is pushed in by water pressure. As the Water scorpion heads to the surface, the air in the lower pits will be compressed more than that in pits nearer the head, so the insect receives information about its progress.

True flies of the order Diptera can be distinguished from other insects by having only one pair of wings. The second pair have been reduced to two club-shaped halteres or balancers. During flight these beat in time with the wings and act like gyroscopes. A gyroscope resists change of direction of movement. If the fly changes direction a strain is thrown on the base of each haltere, which is trying to head in the original direction, and the strain is recorded by sense organs.

How the sense organs work

Recent advances in our knowledge of sense organs have shown that they are more complex than was once thought. The eye used to be thought of as being analogous to a camera, with lens, iris diaphragm and light-sensitive film at the back. This is an over-simplification. Behind the layer of sense cells in the retina, there is a complex network of nerve fibres, linking the sense cells with the nerves leading to the brain. Here information gathered by the sense cells is filtered, analyzed and coded into messages to be sent to the brain. Some form of processing is needed in the sense organs to prevent the brain being overloaded. This is particularly

the case in insects and other invertebrates in which the brain is small and often limited physically by the size of the animal. The environment is so complex that the amount of sound, light, chemical and other modalities bombarding an animal would overwhelm its nervous system unless irrelevant information was filtered out and the useful information simplified.

The filtering and transformation of information This process can be effected either at the sense organs or in the brain itself. So effective is it that we often find that an animal's behaviour is triggered by very simple stimuli. A single substance, bombycol, landing on the antennae is all that is needed to make a male silkmoth start searching for a mate. Filtering in the sense organ takes place at a variety of levels. The human eye is sensitive to light only in what we call the visible spectrum — ultraviolet and infrared light make no impression — and to light above a certain level of intensity. The lens focuses light from objects in front of the eyes so that we concentrate on a narrow field of view. Within the retina and its associated nerve network there is a further selection of information, which transforms a straight-forward image into a series of messages.

Herons shade their eyes
Herons catch fish, trad-itionally standing patiently in water to seize their prey with a quick jab of the dagger-like' bill. They also use other tactics. A few species dance a pirouette which disturbs the fish and brings them into the open. Others have the trick of opening one wing and shutting it quickly. The sudden shadow makes the fish move and, as likely as not, swim towards the waiting bill. One species counteracts the glare of the sun on water, or seems to be doing so, by turning its back to the sun, spreading a wing and putting its head under it.

It is this last tactic which has been taken to extreme by the African black heron. It spreads its wings, then brings them forward to form a canopy, tucks its head under the canopy and starts to feed as soon as the small fishes retreat into its shade, picking up one after another. In this attitude it looks like an umbrella moving over the water, a curious sight when thirty or forty Black herons are fish-ing in one part of a lake.

Since the Black heron also carefully keeps its back to the sun, the whole pro-cedure looks like a neat de-vice for avoiding glare as well as calling in the fish to shady shelter.

It is, however, an innate action, not intelligent, since the bird does the same on days when clouds hide the sun, and will, moreover, continue to fish like this even at night, which surely must be no more helpful than the ostrich's alleged habit of sticking its head in the sand.

The working of the retina has been studied in the frog, an animal whose behaviour is dominated by reflexes that give it virtually no flexibility of behaviour. The organization of the nerve connections between eye and brain virtually does the frog's thinking for it. They select what is important in the frog's environment and the frog acts accordingly. The retina of the frog's eye is made up of three kinds of cells. The sense cells, which are stimulated by light falling on them, are connected to bipolar cells which connect with a number of ganglion cells, whose fibres make up the nerve trunk leading to the brain. There are four sorts of ganglion cell, each with a special analyzing function. One sort reacts to sharp contrasts of light and dark and acts as an 'edge detector'. It serves to delineate the outlines of the scenery confronting the frog — curves of boulders and outlines of trees. The second responds to darkened parts of the scene and presumably helps to show up hollows and shady places. The third and fourth react to movement and to small, round objects respectively and are known as 'event' and 'bug' detectors. These two seem suited to help the frog find its prey and this is supported by evidence from a study of the frog's diet. Frogs eat a wide variety of insects and other small invertebrates, but the proportion of predatory animals in their prey is much higher than would be likely if the frogs collected them at random. This is because the frogs react specifically to small moving objects, and predatory animals necessarily move about more than plant-eating animals. Spiders and beetles are the commonest prey of the frogs, followed by active flies and caterpillars. Butterflies, which settle and sit still, are rarely taken, except by species of frog that jump up and catch them as they fly low.

The 'event' detector is also useful for notifying the frog of approaching enemies; its reaction is to leap into the safety of a pond or stream. Here, it is helped by another filtering device. Frogs have colour vision and are specifically sensitive to blue light. As water reflects blue light strongly, even on a cloudy day, any sheet of water will act as a beacon, guiding the frog as it leaps to safety.

Despite the peripheral filters in the sense organs, the brain is still receiving a large amount of information, most of which is not needed at any one time. The brain selects certain information as being relevant to a particular behaviour pattern. This is **selective response**. Whereas we would take in the whole of a situation, an animal's instinctive behaviour can be triggered by one stimulus. A small, rounded moving

A butterfly landing on someone's finger tastes the small amount of sweat on it through its feet and the tongue immediately uncoils to imbibe the liquid.

object is the stimulus that triggers the feeding response of a frog. The frog does not take into account the detailed shape or colour of the insect, unless it has learned to avoid brightly coloured poisonous insects. Such a simple mechanism is economical in terms of the nerve cells needed in its operation and in its speedy triggering of hunting behaviour, but its rigidity can lead to mistakes. Frogs sometimes eat falling leaves — or even small frogs of their own species.

How stimuli evoke behaviour

We have already come across a variety of stimuli. These are the messages from the environment that act, via the sense organs, on the nervous system. They provide the animal with information, without which it could be helpless. Stimuli affect behaviour in three ways. First, they arouse the animal and provide it with the necessary motivation for behaviour. An example is the female canary having her breeding behaviour 'primed' by stimuli from the male, and from the nest, acting on her sex hormones (p 25). The continuous stimulation of nest and eggs keeps her incubation drive active throughout the egg period. At a simpler level, light falling on the eyes of many insects has a stimulatory function. Some butterflies will stop flying and fall to the ground if the light is switched off and cockroaches are more responsive to other stimuli when exposed to light. We have already discussed the second way in which stimuli can affect behaviour, by eliciting responses. This can be seen in simple reflex action as well as in complex levels of behaviour. Once having started a pattern of behaviour, stimuli may then proceed to guide it; this is the third way in which stimuli affect behaviour.

A pattern of instinctive behaviour is often elicited by a single stimulus or set of stimuli from the environment. This is called a **sign stimulus** or **releaser** and it can be 'dissected' from the plethora of stimuli that the animal is receiving. In a classic experiment, David Lack showed that a European robin redbreast will attack a bunch of red feathers wired to a branch. The red feathers are the sign stimulus that releases the robin's aggression. A complete robin is not necessary; a red mass alone conveys the message that an intruder needs to be attacked. In North America, the male Red-winged blackbird bears scarlet epaulettes on an otherwise plain black plumage. Each male sets up a territory in early spring and, if confronted by another male, it displays by showing off its epaulettes. Once the territory is established, each blackbird

Switching behaviour

The most familiar butterflies of the northern hemisphere are the 'whites' of the family Pieridae. They are abundant and conspicuous but, above all, they are garden pests. The Large white, as well as the Small white, which has been introduced from Europe to North America, lay their eggs on the leaves of plants of the cabbage and mustard family, Cruciferae. The caterpillars proceed to strip the leaves and ruin the crop.

The butterflies choose a suitable place for egg-laying by drumming with their forelegs. During the search, they drum only on green or bluish-green surfaces, that is the colour of the leaves their larvae prefer. But before they lay their eggs, they must feed; they are stimulated to make feeding movements with their long, tubular tongue by the colours of flowers: red, violet and particularly yellow and blue. At this stage they are indifferent to green. Male whites are also attracted to court butterflies of other species if these have yellowish-white wings. So the behaviour of white butterflies is mediated by colour vision. There is some internal mechanism, perhaps hormonal, that switches on selective responses at the appropriate times.

Only female mosquitoes puncture our skin to suck our blood. To a female mosquito that drink of blood is truly important, because without it her eggs cannot mature. She finds the host from which to suck, whether it be a human being or other warm-blooded vertebrate, by a sequence of reactions to the environment.

A resting female is stimulated to take wing when she detects a small rise in the concentration of carbon dioxide in the air around her, such as would be supplied by the exhaled breath. Once airborne, she moves more or less at random but with a tendency to fly upwind. Sooner or later she enters a zone of warmer and wetter air, such as would also come downwind from a living animal. Once in this host stream she continues to fly upwind; but should she fly out of the stream, her immediate response is to turn into it again and then turn again upwind. She determines her course upwind, not by wind pressure, but by the apparent speed of movement of objects beneath and to the side as she passes them. The nearer she gets to her intended victim, the warmer and more humid the host airstream becomes: the concentration of carbon dioxide also rises but it is temperature and humidity that are important.

The object of mosquito repellents is to mask the initial stimulus that the carbon dioxide in our breath provides to the mosquito. A person standing in still air is in a column of convection currents rising from the body. Because this convection column is continually rising, it is true to say that an insect repellent on the sock will protect the face as effectively as a repellent placed on the face itself.

mates with up to six females who build nests within the territory. If the epaulettes are obscured with a dye at the beginning of the breeding season, the male has no difficulty in attracting females and fathering broods but may have difficulty in holding his territory. He is likely to be driven away by his neighbours. So, epaulettes are releasers in a territorial, not a sexual context.

Sign stimuli can be found throughout the animal kingdom and those that come from other animals rather than the inanimate environment have often been evolved as special signals. The red feathers are a releaser of a robin's aggressive behaviour. Releasers are not confined to signalling between members of one species. The black and yellow stripes of a wasp could be called a negative releaser because they stop a bird from attacking, warning it that the wasp is dangerous.

The action of releasers may be modified by the physiological state of the animal. A releaser may even disappear when there is not the correct hormonal background. Male Three-spined sticklebacks defend territories during the breeding season. They develop a red coloration on the belly which elicits aggression if they stray into a neighbouring territory. Female sticklebacks have no red and are allowed into the territory. Outside the breeding season, the levels of sex hormone subside, the red coloration disappears and male sticklebacks gather peaceably in shoals.

The response to a stimulus is often related to the strength of the stimulus, although this is not always the case. An antelope runs no harder when pursued by two lions than when pursued by one. Yet a stronger stimulus may evoke a strong response or direct the response away from a weaker stimulus. By presenting models of stimulating objects ethologists have found that some models represent a **supernormal stimulus** that evokes a stronger response than does the natural stimulus. Thus, an oystercatcher will leave its own eggs and attempt to incubate a model egg almost its own size.

If some artificial stimuli evoke stronger responses than do natural ones, one may ask why sign stimuli have not been evolved to produce this stronger response. The answer is that an animal's behaviour is made up of compromises. A huge egg will never be ignored by any oystercatcher with the slightest drive to incubate but in practice the bird could not possibly supply such an egg with the necessary warmth. Neither could the bird lay such a monster. Super-efficient signalling has to come second to more practical

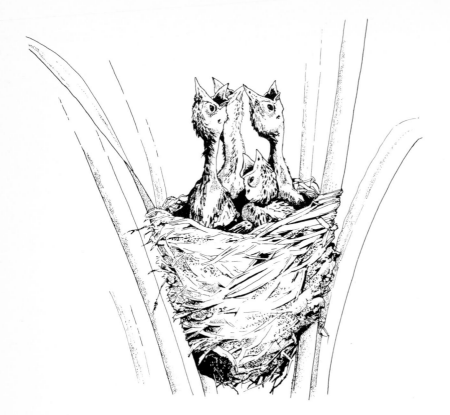

Nestling Reed warblers greet the parent returning with food by gaping in its direction. The hungrier they are the more persistently they gape.

considerations. Because of the compromise situation, supernormal stimuli are mainly experimental oddities but a few occur naturally. The chicks of many small birds **solicit** food by gaping. They present a widely-opened bill to show a brightly coloured mouth, which acts as a releaser causing the parent to drop food into it. The huge mouth of a baby cuckoo acts as a supernormal stimulus so that passing birds will feed the cuckoo rather than carry the food to their own nestlings.

The role of the effectors

Stimuli from the environment, hormones within the body and half-understood processes within the central nervous system combine to regulate an animal's behaviour. But what makes up this behaviour? It consists of sequences of muscle contractions that make up movements as varied and ordered as the stimuli and nervous co-ordination that trigger them. The ultimate function of an animal's behaviour is to adapt to its environment. The senses inform it about the environment and the effectors perform the necessary adaptations. By effectors, we generally mean the limbs that move the animal, the tail that helps to balance or is used in swimming, the jaws

When is an egg?

For a gull, an egg is a different thing at different times. An egg in its own nest is to be nurtured while an egg in another bird's nest is food. Furthermore, when its own eggs have hatched, it has to remove the empty shells because their white lining will attract predators. The problem of how to treat a particular egg is resolved by a number of **sign stimuli**. When searching for food, a gull recognizes eggs by their shape, but shape is not an important feature of eggs that need to be incubated. By testing what sort of model eggs are rolled into the nest (p 160), Dutch ethologists have shown that the speckling of the eggshell is the important stimulus for releasing, retrieving and incubating behaviour. The more speckles and the greater their contrast with the background colour, the stronger is the urge to incubate the egg. When the chick has emerged, the parent gull is stimulated to carry the shell away by the sight of the ragged edge and the white border formed by the membrane lining the shell. If, however, the shell is heavy, the gull releases it as this indicates that, although the lid is off, the chick is still inside.

A gull is not fussy about what it incubates, provided the above criteria are fulfilled; but guillemots incubate only their own eggs. They do not make nests but pack together on cliff ledges, their eggs balanced under them. If an egg is displaced, its parents alone will retrieve it, having recognized its patterning.

and other mouthparts that are used in feeding. Limbs and jaws are also used in defence, as are stings. Feathers and hair can be fluffed up to help keep the body warm. Special pigment cells change shape to alter the animal's colour. All are effectors.

In the same way as sense organs are analyzed to find out how they code their messages to the brain, so are effectors studied to see how the complicated movements of, for example, a salmon leaping a weir, or a horse galloping across a field, are made up of individual muscle contractions. Here, however, we are more concerned with movements which are used as signals between animals. Because signals are stereotyped, they are relatively easy to analyze to give an idea of the nervous mechanism that controls them.

A flock of starlings roosting on a tree suddenly flies with a whirr of wings. They move as a solid mass and, when airborne, the flock wheels and soars as one bird. A platoon of soldiers can show such precision, but only in response to shouted commands. The commands to the starling flock are **intention movements**. As a bird prepares to take off, it flexes its legs and half opens its wings ready to launch itself into the air. The roosting starlings are all the time keeping a close watch on each other so that as one prepares to take off, the others follow suit and there is a co-ordinated spring skywards. Intention movements abound in everyday life. They can be best illustrated by the small involuntary movements that we make when we are slightly hesitant about making a move. The individual starling is hesitant about taking off but similar movements by its fellows reinforce its urge and precipitate action. Similarly, once airborne, if one starling shows the intention to wheel to the right, say, these movements are seen by the rest, who immediately follow suit. They do so with such speed that all seem to be turning simultaneously.

In the course of time, many intention movements have been modified to form more precise social signals. The process is called **ritualization**. An aggressive Herring gull stands with its wings held 'akimbo'. They are held slightly from the body, an intention movement for flying that has become ritualized into a signal that the gull is liable to launch itself at an opponent. A feature of ritualization is that the actions are often distorted so that they bear little resemblance to the original.

Signals used in aggression and courtship have also been formed by ritualization of **conflict behaviour**. When an

55

animal is confronted by an opponent, two urges come into conflict: the drive to attack and the drive to flee. Its ultimate behaviour depends on the resultant strengths of the two drives and even if neither drive is strong enough to make the animal attack or flee, the underlying motivation will show in its actions. In a sexual context, a third drive is added. When a male animal first approaches a female it is torn between desires to attack, to flee and to mate. Courtship displays are the resultants of the three conflicting drives.

A conflict situation often leads to strange behaviour. In man, an inner emotional conflict leads to such irrelevant behaviour as scratching the head or chewing the lip. These are displacement activities (see p 25), produced when feelings cannot be expressed properly. Under the same circumstances, a bird preens itself and 'displacement preening', as it is called, has become ritualized to form a variety of courtship displays. Zebra finches, popular cage-birds, bow during courtship but the bow is really a ritualized bill-wiping movement, an irrelevant comfort action like preening or scratching.

Another expression of conflict is the **redirected response**.

Grey heron parents take turns on the nest. The changeover is preceded by a statuesque, un-hurried, ritualized greeting ceremony.

Blown-up toads
The European common toad has an arch-enemy in the non-venomous Grass snake. Against this snake the toad is defenceless except that it can blow up its body and rise on stiff legs, looking absurdly like a gro-tesque balloon on stilts.
In theory, the toad's snake-reaction should preseve it for it is almost instant-aneous. Moreover, as if it were taking no chances, the toad will inflate when any horizontally moving, slender body passes across its field of vision. Even a grass stem plucked and passed longitudinally near its eye will set it off.

When a supposed danger has passed the toad slowly deflates, but will immediately inflate if the slender object is once more passed across its field of vision. If this is repeated several times in succession, with only a short interval of time between each, the inflation becomes less pronounced with each repetition, either from fatigue or, more probably, through habituation. So, if a Grass snake is hungry, and consequently persistent, there comes a time when its victim gets used to seeing it, with fatal results.

If the animal is very keen to attack, but is also very frightened, it aims its thwarted aggression against a substitute. When facing each other across a territory boundary, Herring gulls display their thwarted aggression by tearing tufts of grass. Nikita Krushchev, in a United Nations meeting, hammered the table with his shoe as redirected aggression against fellow members of the United Nations. It is not such an outlandish idea, that a leading statesman in the highest political gathering should behave like a bird defending its territory. In our emotions we still behave as animals. Not only are the emotions the same — fear, anger — they are controlled and have their expression in the same manner. The concern of civilization has been to learn to control those impulses but they cannot be erased.

Chapter Three
Eating and Drinking

Every living thing feeds. Only animals eat. Plants make their own food with the aid of sunlight, and thereby support all animal life — all flesh is grass. In eating, food must be taken into the body. That is, it must be swallowed. There are some animals that do not eat; they imbibe their food through the skin. These are mainly the internal **parasites,** like the tapeworms, that live in the digestive tract of another animal, bathed in nutriment almost from birth to death.

Chewing and swallowing are the ways of consuming food most familiar to us. Large numbers of animals, especially those lower in the scale, have ways of eating that bypass the labour of chewing, or of gulping. Others swim through food, like whalebone whales and herrings, walk through it, like blowfly maggots, or fly through it as with swifts, swallows and nightjars that fly through swarms of insect prey. Others are the **blood-suckers,** like mosquitoes and Vampire bats.

As a rule, in order to be swallowed food must be cut up or chewed. Sometimes it is swallowed whole. An owl catching a mouse or a small bird swallows it whole and later regurgitates (or vomits) the indigestible parts in the form of a pellet, a rounded mass of hair or feathers and bones.

Between a predator and its prey there is often stratagem and counter-stratagem. A hunting owl's flight is almost soundless; but in the deserts of south-western United States (see p 63) the Kangaroo rat's sensitive hearing can detect even the slight sounds made in time to take evasive action.

Browsers share the trees
'All flesh is grass' is a biblical quotation widely used in ecological texts, but many of the large plant-eating animals prefer the leaves of trees and shrubs to herbage. These animals are browsers, grazing being a term restricted to the eating of grass. Elephants, giraffes, rhinoceroses and many of the antelopes are browsers, preferring the bush country of Africa where, in the denser parts, there may be no grass at all. Some browsers, like the rhinoceroses, are coarse feeders, taking in twigs as well as leaves and even consuming bitter thorn apples. Others feed delicately. The antelopes have slender muzzles that allow them to nibble leaves of acacia trees without being spiked by the wicked thorns. The giraffe has a long, prehensile tongue that stretches between the thorns.

Although there are many browsers in the African bush, they avoid competing with one another by choosing different food supplies. Giraffes feed on the high foliage up to 18ft and they distort the shapes of trees, flattening the tops of low trees and pruning taller ones into oval or hourglass shapes. The gerenuk, a graceful long-necked antelope, can reach 6 ft by standing on its hind legs, but the tiny dik-diks feed only at the lowest levels. Elephants spoil the feeding for everyone — when they cannot reach the tops of trees they knock them over. In this way, they have turned areas of bush country into open savannah.

Another important group of animal feeders are the **scavengers,** including vultures and hyaenas, which feed on carrion (dead animals). They have their counterparts in the **composters,** such as earthworms and soil insects, that feed on dead plant materials, such as dead leaves, converting them to humus. A further group are the **omnivores** (eaters of everything) that eat plant and animal food more or less indiscriminately. The pig is a familiar example, the ostrich is by tradition another. They will eat almost anything except broken glass — the North American porcupine will do even that, gnawing glass bottles occasionally.

Virtually every organic product is used as food by one member or another of the animal kingdom. Therefore, food in its broad sense includes every kind of plant or animal, living, dead or decaying, blood, bone and tissues and all the by-products of plants or animals, including dung (eaten by **coprophages),** as well as bacteria. Finally, there is **cannibalism,** the eating of one's own species. Gulls will kill and eat a wounded gull and a surprising number of lions, especially cubs, are eaten by other lions.

Most animals can be broadly divided into **phytophages** (Greek for plant eaters) and **carnivores.** Phytophages may be **browsers,** that eat leaves and twigs, or **grazers,** that eat grass almost exclusively. In addition there are **fruit-eaters** and **leaf-eaters,** and those that feed exclusively, or almost exclusively, on nectar or nectar and pollen, like the honeybee.

Phytophages One problem for phytophages is that plants are not easily digestible. Digestive juices cannot deal with the cellulose cell walls of plants so these must be ruptured by chewing. As a consequence, grazing and browsing animals have large blunt teeth for crushing plants and must eat slowly. Cattle and other cloven-hoofed animals reduce the time that they spend in the open, where they are exposed to predators, by **rumination.** Having filled the rumen, the first compartment of their four-chambered stomach, with barely-chewed food, they retire to the safety of cover and ruminate or 'chew the cud'. They bring up one bolus of food at a time into the mouth and masticate it thoroughly before returning it to the rumen and thence to the other compartments for digestion. The cellulose is digested by bacteria and Protozoa which multiply and are themselves later digested by the ruminant animal.

A similar procedure is used by rabbits, hares, shrews and some rodents such as the House mouse and Common rat.

This is **refection,** in which certain droppings are eaten so that food passes through the digestive tract twice. Young rabbits eat their mother's droppings to obtain the necessary cellulose-digesting organisms, known as the **intestinal flora.**

This eating of the maternal excreta, known also as coprophagy, is most pronounced in the koala, of Australia. A month before the baby, still in its pouch, can start eating eucalyptus leaves, it eats straight from the mother's anus, regularly each day between 1500 and 1600 hours. What is taken is half-digested leaves, so it probably receives nourishment as well as stocking itself with an intestinal flora.

The African ground squirrel and several hoofed animals are known to eat earth when young before they start to take solid food. It is believed that this is yet another way in which young animals introduce the necessary micro-organisms into the gut to form an intestinal flora.

One of the most remarkable feeders is not strictly a phytophage, yet it depends utterly on plants to live. It is a tiny flatworm, *Convoluta*, about a quarter of an inch long. Most flatworms are carnivorous. They live in fresh water and feed on small freshwater worms, crustaceans and small freshwater snails. Mostly they are about an inch long, although one living in Lake Baikal in Soviet Asia is ten inches long. *Convoluta roscoffensis* lives buried in the sand on the seashore. It has unicellular plants (algae) living in its tissues. When the tide is out by day it emerges from the sand so that light can penetrate its tissues to allow the green algae to manufacture starch by the aid of sunlight. *Convoluta roscoffensis* lives exclusively on this starch.

Hunters: active and passive   The name 'carnivore' should strictly be reserved for animals feeding on red flesh but it tends to be extended to include those eating others animals, of any kind. Non-herbivore is a better word. For the most part non-herbivores must chase after and capture food or must have specialized devices for trapping and subduing it, or else they must capture it as it passes by or attract it to themselves. Some use a combination of two or more of these methods. Examples are all around us. In the first category are the hunters, pure and simple, like the lion, tiger, wolf and coyote. Then there are the many birds-of-prey, such as eagles, hawks, falcons and owls, as well as the many kinds of insectivorous birds.

The hunters must outwit or out-smart their prey if they are to survive. So we find they are generally more intelligent than the herbivores, although intelligence is not necessarily a

The lion's share
The male lion with his magnificent mane, the King of Beasts, is a lazy animal. Nine out of ten kills made by a pride of lions are the work of the lionesses. In a communal hunt, the males tend to hang back until the kill is made, then they come forward to claim the carcase. The lionesses and cubs have to wait for their meal until the lions are satisfied. As far as hunting goes, however, the lionesses are better off without the lions. The latter are slower on their feet and their large size and bristling mane make them conspicuous.

When making a kill, a lion first knocks the prey over with a blow of the paws, and smaller animals are killed almost immediately with a bite on the nape or throat. Larger animals are more difficult to deal with and first they must be dragged over. The lion then seizes the throat or muzzle. If it is held by the throat, the prey dies by strangulation but if the lion places its mouth over the prey's muzzle, death is by suffocation. In both cases, death comes slowly but the animal does not struggle as, mercifully, it seems to be in a state of shock (p 272).

A she-bear suckling her young. The ability to supply milk for nourishing their offspring is one of the principal characteristics of mammals.

monopoly of the hunter. Outstanding examples are the fox, which must be more crafty and resourceful than the rabbit it hunts, and the weasel, which has more brain power than the mouse. Nevertheless, hunters frequently miss their prey. Lions and tigers are only 5-10% successful, African hunting dogs 85% successful.

Between a hunter and its prey there is often stratagem and counter-stratagem. The hunter stalks its prey with the utmost care to keep concealed and silent but it has to rely largely on surprise to outwit its prey's alert hearing and sharp vision. In the arid areas of the south-western United States the Kangaroo rat lives in danger of night hunters such as owls and snakes. Owls hunt by vision and hearing, both of which are incredibly acute, and rattlesnakes hunt by means of their facial pits (p 47). The Kangaroo rat's safety depends on its superlative hearing which is tuned to be particularly sensitive to the faint susurrations of an owl's muffled wings or the scratch of a snake's scales on the sand. It is said to have **tuned ears.** Once alerted it jumps clear. The hunters must rely on surprise or on catching the inexperienced, the sick or the injured. In practice, predators mainly weed out the unfit.

Attack and countermeasure are even more displayed in the bats. The majority of bats hunt insects at night by means of echo-location (p 43), emitting streams of **ultrasonic** clicks of incredible intensity. The clicks of the Little brown bat measure 60 dynes per cm³, as compared with 3 dynes per cm³ near a pneumatic drill. Insects are detected by listening for returning echoes but the exact mechanism varies between bats and is extremely sophisticated. Little brown bats can find Fruit flies at least twenty inches away. To combat this detection system, some insects such as moths and lacewings have 'ears' that can pick up the bats' clicks and they take evasive action by dropping to the ground. One group of moths even defend themselves actively. They emit their own ultrasonic pulses which advertise that they are distasteful.

By contrast with the active hunters, frogs and toads capture living prey by sitting and waiting for insects to pass their way, although they also seek them out. A frog's brain is a simple affair but the animal makes up for this by doing some of its 'thinking' with its eyes. When a fox sights a rabbit or a bird, signals (or messages) are conveyed through its eyes to its brain and the information is there processed. In a frog the information is mainly processed in the nerves behind the retina of the eye.

A peculiarity of many hunters is that they must gulp their food after capture. There is often little time for leisurely meals, if the hunter is not to go hungry. But before this, the prey must be seized. A dog seizes its prey in its jaws and uses its teeth to kill. A starfish takes hold of a mussel with the tube-feet on the undersurfaces of its arms and forces the shells apart. Then instead of pushing the body of the mussel into its stomach, it everts its stomach over the mussel.

Snakes are specialists at seizing, as befits animals with no limbs. The most primitive snakes are the boas, the family that also includes pythons and the anaconda. These kill by constriction, not squashing the victim to a pulp as is often supposed, but by throwing tight coils around its body so that it cannot breathe and dies of suffocation. Other snakes subdue their victims with poison which is injected through fangs — modified teeth. In the Pit vipers and rattlesnakes the fangs are folded back out of the way and are erected when about to be used. The potential danger of a venomous snake depends not only on the strength of the venom but on the amount injected and its mode of action.

The snake's victim is swallowed whole and large prey can be taken in through the loose attachment of the jaws allowing the throat to be distended. The jaws work alternately, pulling at the prey to draw it into the mouth.

Food-passing
Harriers are long-legged, slender-winged birds of prey and they indulge in a spectacular food-passing ritual. The male calls to the female as he aproaches the nest. She flies up and, after some manoeuvring, the male drops the prey for her to catch in mid-air or, more impressively, she flips upside down and seizes it from his talons. The ritual continues after the eggs have hatched, the female taking the prey back to the nest where she dismembers it and feeds the chicks.
In the Hen harrier or Marsh hawk, when the chicks are two weeks old the female begins to bring more food to the nest than the male. By this time, also, the youngsters have grown more feathers and have become more active. On leaving the nest they remain in the vicinity. They soon learn to respond to the female's call and quickly learn the technique of food-passing.

Foxes have a reputation for intelligent, if not cunning, behaviour. There are many stories of their outwitting hounds and of losing no opportunity to steal a meal. Many foxes now live in towns and suburbs where they have learnt to raid dustbins. This vixen has knocked over the dustbin to remove the lid and get at the contents.

Left: A Thomson's gazelle buck scratching, against a background of Blackwinged stilts and Lesser flamingos. Scratching is a stereotyped behaviour. Every member of a species scratches in the same way and the movements involved constitute a fixed action pattern. No matter what may be done in an attempt to alter it, the scratching cannot be carried out in any way other than that characteristic of the species.

Below: The Greater waterboatman or backswimmer is a predatory bug which seizes other water creatures and sucks their juices. It is guided to prey first by picking up vibrations caused by their movements in the water, then by sight. When the prey has been seized, the backswimmer eats it only if the taste and texture are right.

Right and far right: Perfect co-ordination between eyes and tongue are needed to register a hit. A High-casqued chameleon has slowly crept up on this grasshopper. While taking aim, its mouth opens and the tongue is moved forward into the ready position. Strong muscles contract to flick the tongue out in a lightning flash but this time the chameleon fails because the grasshopper is holding tight to the grass stem. A second attempt is needed to secure it.

Below left and below: Sequence of attack: a Common octopus reaches out with one arm from its home in an empty triton shell to investigate a crab, sampling it by touch and by means of the taste-buds in the sensitive arm-tip. Satisfied with its choice of a meal it oozes out of the shell, pounces on the crab and bites it. While waiting for the poison to subdue the crab, the octopus envelops it then carries it back home. After a while, the completely cleaned out crab shell is 'spat' out with a jet of water.

The Mascarene frog of Africa is beautifully camouflaged as it sits motionless among grass or rushes, which is how it manages to catch a dragonfly in spite of that insect's excellent eyesight. It leaps and grabs with its mouth and tongue fast enough to catch insects flying on the wing. The frog reacts only to moving prey. If the dragonfly were to settle and remain still, the frog would not see it.

A Puff adder swallows a Grass rat. With no limbs to help it cram food into its mouth or to hold the prey down while pieces are torn off, the snake uses its jaws to pull the rat into its mouth. The jaws are loosely attached to the skull and to each other so that they can open wide to accommodate large prey and work alternately to draw it in.

All animals harbour parasites. Some parasites are merely irritating, others carry disease. This lioness seems not to be worried by an infestation of blood-sucking insects feeding on her neck; the little damage they do can best be assessed by her yawn. The truth is her yawn is not one of boredom or distress; she has awakened after a day of dozing in the shade. The yawn fills her lungs deeply, drives extra oxygen by way of the blood to the heart, whence it is pumped to the muscles. The muscles become ready for action, and tiredness is banished. Lions also yawn in the heat of the day in order to cool themselves by increased evaporation. (See also Chapter 8.)

Devices for capturing prey   Many non-herbivores make use of special devices to capture prey as it passes. The waving tentacles of a Sea anemone and the spider's web, even the extraordinary tongue of a chameleon, are simple and familiar examples of this, although chameleons also stalk their prey to some extent. An alternative often encountered, which also conserves energy that would otherwise be spent on hunting, is to attract prey to the predator, either by a **lure** or by some trick of behaviour. Deep-sea fishes, living in perpetual darkness, where prey animals are few and far between, use lights as lures to attract prey to them. Usually they also have a huge mouth not only to make sure of their capture but also to be able to cope with prey of almost any size. Their relatives in shallower waters, such as the Fishing frog and anglerfishes use similar lures that are non-luminous.

These are only a few of the specialized feeding habits. Other animals utilize specialized habitats, like the koala and the termites, or use specialized **feeding associations,** either habitually, as in Cattle egrets (p 228), or occasionally, as in fishes known as puffers and scats (p 226). Vultures rely almost entirely on others to capture and kill prey for them.

Gluttony   The wolverine has earned the alternative name of glutton but there are other contenders for the title. One, the European Common toad, was once described as eating any living animal capable of being swallowed, including insects and their larvae, especially beetles and ants, woodlice (sowbugs), worms, snails, young newts, frogs, toads and lizards, and newly hatched Grass snakes. Ironically, the adult Grass snake is the toad's main enemy.

One European Common toad was watched sitting beside a beehive picking off the bees one by one as they returned. Of toads of the same species, when tested, some would swallow bees, others would not touch them. Another toad when dissected had 363 ants in its stomach, and yet another, confronted with a liberal supply of earthworms, ate so many in rapid succession that in the end it passed the worms alive.

Almost all toads, and frogs, catch smaller prey by flipping out the tongue which is, unusually, fixed to the floor of the mouth in front. There it lies coiled to be flipped out like lightning. The tongue is always said to be sticky, and doubtless saliva alone has adhesive qualities, but the advent of high-speed photography revealed a different story. About twenty years ago, an American naturalist used an electronic flash to photograph a bullfrog feeding. He used mealworms each held in the light grip of a small spring clamp. The frog had no difficulty in pulling a mealworm from the clamp, which would not have been possible if only stickiness were used. The photographs taken showed the tip of the tongue wrapped round the mealworm.

Gluttony may be necessary to take advantage of moments of plenty to offset periods of scarcity. That it occurs in many animals besides those quoted also goes to show that animals do not always know when they have had enough. This is true for the lowest to the highest. Even so, gluttony is not a normal feature because satiation causes an attenuation of the feeding drive.

A Sea anemone is one of the lowest forms of life. It is no more than a fleshy bag with only one opening, the mouth, surrounded by a crown of tentacles armed with myriad stinging cells. A small fish touching a tentacle is stung, paralyzed and held. At the same time the tentacle begins to curve over towards the mouth. This action stimulates neighbouring tentacles to do likewise and together they stuff the prey into the mouth.

Herring gulls following the plough will sometimes go on gobbling the turned-up earthworms until they can eat no more, then fly off to vomit up the worms and return again to the feast.

Squirrels are usually given the reputation of being the archhoarders, yet some species of mice and rats are even better than they are. Indeed, many times stores of nuts and acorns have been unearthed and have been supposed to be the work of squirrels when in reality they were hoards brought together by mice.

The pet lover knows that hamsters will stow away any surplus food in special pouches in their cheeks. In the wild they raid crops for the same purpose and carry the food back to their burrows. These pouches can stretch to an enormous size. The Pocket gopher of North America is so-called for the enormous cheek pouches it uses to the same end as the hamster.

The Mole rat of the Old World eats bulbs, and it stores them, first biting out the growing point of each bulb, so that it cannot sprout but remains fresh and juicy for a long time. The European mole eats mainly earthworms. When earthworms are very abundant, a mole may hoard them. It first bites off the tip of a worm's head, then ties its body in a knot, and then stores these incapacitated animals in chambers underground. In time each worm grows a new head and wriggles free of the knot, but should the mole need extra food before that happens it can take it from its living cache.

Any small animal touching a tentacle will set this off, and the experiment was once tried to see how much food a small anemone would swallow. Small pieces of flesh from a dead prawn were prepared and one after another quickly placed on one tentacle after another. All the tentacles curved over to the mouth and all the prawn meat was swallowed by the now bloated anemone. After a short time the anemone turned itself almost inside-out. The tentacles moved slowly down the body and the mouth disc became elongated until at last the tentacles were at the base of the body as the bolus of prawn meat was ejected through the mouth. Then the anemone lay on its side, limp as if exhausted, before slowly resuming its normal shape. After this it was ready to feed again.

Herring gulls will come inland and follow the plough, gobbling up the earthworms exposed as the earth is turned. From time to time one of them will fly away, vomit a cropful of worms and return to the feast, recalling the practice of the Ancient Romans with their vomitaria, the use of which enabled guests at a banquet to taste a wide variety of dishes.

Feeding frenzy   Disgusting in human beings, this vomiting may be an occasional practice for more animals than are at present suspected. It may be the result of a **feeding frenzy,** a term usually reserved for a behavioural trait of sharks. When one shark makes a kill the smell of blood diffusing through the water attracts other sharks to join the feast. Excitement mounts until the sharks, seemingly crazed, turn and rend each other.

It is well known that if a fox gets into a chicken house it will kill many chickens and take only one away to eat. The reason for this is not immediately obvious. One way is to regard the fox as a bloodthirsty killer. Another is to suppose the poultry panic, and that this excites the fox's hunting instinct so it goes on killing. It looks like a form of feeding frenzy, for a fox has been known to kill as many as seventy birds in such an incident. There is, however, a more sober explanation (see **overkill,** p 268).

The Red fox, the traditional despoiler of chicken houses, has a good appetite but does not over-eat. It buries surplus food, as a dog buries a bone, returning to it when necessary. A fox has been known to locate and dig up carrion buried by man to a depth of nearly two feet, so re-finding its own caches can present few problems.

Eating and over-eating   Presumably a fox must have a safeguard against over-eating similar to that of the Common rat, which was investigated some years ago. It was found that the amount of food taken by the rat was controlled by the pituitary, a small gland on the underside of the brain. In the stalk connecting the pituitary with the brain is a control centre. If this is injured a rat will go on eating long after its hunger is assuaged. Such a rat may double its weight in eight weeks.

Beside this control centre is another that stimulates the animal to eat. When this is damaged a rat will refuse food and soon become emaciated. These two centres, each consisting of a few cells only, act like a thermostat, one cutting in to make the rat feed, the other switching on to make it stop eating. The effect of the central control of feeding is to make the animal feed in bouts. A rat usually eats every two to four hours but some animals cannot feed in bouts. Male Fur seals do not feed at all while they are defending territories in the breeding season. They have to subsist on their stores of fat and become very emaciated. After that they must feed heavily to replenish the stores of fat. The same holds for Emperor penguins that incubate their eggs for several weeks without relief.

More studies have been made with laboratory rats than any other animal. German scientists found that rats can survive for ten to fourteen days without food. Their weight is halved and many body tissues degenerate, although the brain does not. At Cambridge University, it was found that young rats given a super-abundance of food, so that they never go hungry, are more active and more inquisitive and learn more rapidly, than rats given adequate food and no more. By contrast, rats compelled to live on an adequate but frugal diet live longer than those that consistently gorge themselves. They also remain youthful; a two-year-old looking as youthful as a three-month-old that has fed well. There would seem to be a moral here for the over-eating members of some human communities.

In books on natural history written in the first half of this century there are frequent references to this or that animal eating more than its own weight in food a day. This applies especially to some of the smallest mammals, such as shrews and mice. In the last twenty years, careful studies have produced more sober results, which have a distinct bearing on the supposed gluttony of some species. Thus, the earlier naturalists may have caught shrews and taken them home to keep under observation in captivity. On giving them food, on arrival home, they were astonished to see how much the

First things first
There is a simple, if ribald, story of a visitor to a zoo seeing two monkeys on a perch high up in a cage. He offered them a banana. The monkeys ignored it. The visitor asked a passing keeper why the monkeys showed no interest in their favourite food. The keeper, more versed in the ways of monkeys, looked up and saw they were making love. He remarked: 'If I were one of those monkeys, I wouldn't come down for a ship's cargo of bananas.'
If nothing else, this story illuminates the result of a very recent piece of research: that animals have a **hierarchy of behaviour.** They are often exposed to mixed, even contradictory, sensory inputs. It may be impossible to react to all these at once since the responses may be mutually incompatible. So they have a system of priorities. The research was carried out on certain carnivorous marine snails, in which feeding takes precedence over everything else, possibly because their food supply is normally limited. Thus, the response of one of these snails to finding itself on its back is normally to turn over immediately. When it was experimentally laid on its back and food was placed beside it, the snail did not right itself but started eating, and would continue to do so for up to an hour. It also would eat rather than mate, and would even interrupt its mating if food were offered it.
Apparently the top priority, that took precedence even over feeding, was egg-laying — presumably to ensure the snails do not eat their own eggs.

shrews ate. They had overlooked — or perhaps did not know then — that a shrew will die of starvation if kept without food for three hours. Since some time would elapse between capture and the time the animal was fed, after the journey home, the shrew would be ravenous. The truth is a shrew normally eats two-thirds of its body weight a day.

Another source of inaccuracy in such estimates comes from the type of food given. A shrew is an insectivore, normally eating insects, although it also requires a few seeds each day. It will also eat small carrion as well as earthworms. A shrew fed on earthworms will eat more than its own body weight a day, solely because there is a higher percentage of water in a worm's body than in the body of a typical insect.

Liquid feeding There are four main methods of taking in food — solid feeding, liquid feeding, imbibing food through the skin, and filter-feeding. The various ways of taking liquid nourishment are almost endless. There are, for example, the hummingbirds (some 300 species) which hover in front of flowers to sip nectar with a long bitubular tongue inside an elongated slender bill. They need the sugars in the nectar, an energy food, to continue their very active lives. They also need protein, and for this they eat small insects and spiders. Their taking of nectar resembles that of night-flying moths, but whereas the moths unfurl the long coiled tongue and insert it in a flower to suck continuously, the tongue of a hummingbird is moved in and out. There are also numerous nectar-feeding bats, in Australia, Malaya, West Africa and South America; they are unrelated yet all agree in having elongated muzzles, a long extensile tongue and weak, almost useless teeth. The tongue is not tubular, as in hummingbirds, but has recurved barbs on the tip for drawing in nectar.

There are some 800,000 species of insects and probably a quarter of these at least take in liquid food. Some make do with nectar alone, since the adult insect does not grow in size and protein food for body growth is not necessary. Then there are the 'suckers', which include those that suck blood, the fleas, lice and bugs, the mosquitoes, gnats and midges, as well as others less familiar. These have tubular mouthparts variously equipped with some form of piercing apparatus. Other insects take plant juices (sap) by the same method as the bloodsuckers. The perfect comparison between the **bloodsuckers** and **sapsuckers** is provided by the troublesome mosquitoes — the female sucks our blood, the male is content with sap. Not all liquid-feeding insects have tubular mouthparts. Some bite and lap the blood or juice, as does the Vampire bat of America, a species of mammal far removed from the vampire of fiction if only by its small size — three inches from nose to tail-tip.

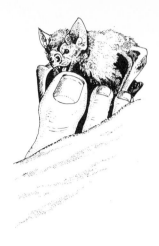

Some of the most habitual 'suckers' among insects are the aphids, variously known as greenfly and blackfly, but collectively labelled plant-lice, in recognition of a similarity to body-lice. Aphids imbibe such quantities of sap that they are constantly dribbling the excess through the anus. This is known as 'honey dew', and where aphids are abundant the foliage and the ground beneath may be coated with a sticky sugary layer. This has led to a benign form of parasitism by ants, which visit the aphids to feed on this exudation.

Living as a parasite   Killing and consuming an animal is known as predation but there are many flesh-eating animals that do not, as a rule, kill the animals they feed off. These are the parasites. Parasitism is an association between two species in which one derives its nutriment from the other, at least in part and sometimes in whole, to the detriment of the other. Viruses, bacteria, certain protozoans, roundworms, tapeworms and flukes are important parasites that cause many human diseases by destroying tissue or liberating toxins into the body. Tapeworms compete for food in the intestine and are not dangerous unless food is short. The Chinese once held that they were even beneficial as two meals could be enjoyed — one sustaining the man and the other the worms. Some arthropods are also dangerous parasites. Blood-sucking fleas, ticks, Tsetse flies, mosquitoes and others do little harm from the amount of blood sucked but the saliva that is pumped into the wound carries protozoans that cause malaria, sleeping sickness and other diseases.

A feature of the parasitic way of life is its complication. Many parasites undergo a number of stages, often involving more than one host. As it is often only by chance that a parasite is picked up by its host there has to be an enormous reproductive output to offset the high losses and asexual reproduction is often used to produce vast numbers.

A typically complex parasite life cycle is that of the Liver fluke. The adult fluke, which has both male and female reproductive organs, lives in the bile duct of a sheep, a horse or even a human. It lays about twenty thousand eggs per day. They pass into the intestine in the bile and are then voided from the host. The egg hatches into a larva called a miracidium which bores into the body of a water snail living

Vampires
The vampire of European folklore is a rather different creature from the small blood-sucking bat of tropical America, yet the latter is not the most pleasant animal. It feeds on large birds and mammals by slicing a hole in their flesh and sucking up the resulting flow of blood. This Vampire bat has reached pest status in Latin America because of its attacks on domestic stock, which are bitten on the ear or in the cleft behind the hoof.
The bat spends some time selecting a suitable place to bite, then licks the area before shaving off the hair and the outer layer of skin with its razor sharp upper incisor teeth. When the blood vessels in the skin are exposed a 'divot' of flesh is cut out and the blood is sucked through two tubes formed by the grooved tongue and lower lip. Clotting is prevented by an anticoagulant saliva and the rasping of the tongue.
Human beings are most often bitten on the big toe, as when sleeping with the feet pushed out of the bedclothes. The bite is not felt and the greatest danger is the transmission of rabies.

at the edge of a pool or on moist ground. The miracidium becomes a sporocyst in which several rediae develop by budding. These burst out of the sporocyst and develop into fifteen to twenty cercariae larvae. Each cercaria swims out of the snail and finds a place where it encysts by forming a waterproof coat. If it encysts on a blade of grass it may be eaten eventually by a grazing animal. The host's digestive juices dissolve the cyst and the young fluke bores through the intestine into the blood stream which carries it to the liver. Here it burrows through the bile duct and the cycle is completed.

Parasitism is by no means confined to disease-causing or disease-producing animals. There are a few parasitic vertebrates, including birds such as the European cuckoo that lay their eggs in other birds' nests (p 159). Skuas, some gulls and the frigatebirds supplement their diet by stealing food from other birds. This is sometimes called piracy but its modern scientific name is **kleptoparasitism.** Great skuas chase gannets, cormorants and terns, harrying them until they are forced to disgorge the food they were bringing in for their young. Similarly, frigatebirds of the tropics harry boobies and tropicbirds.

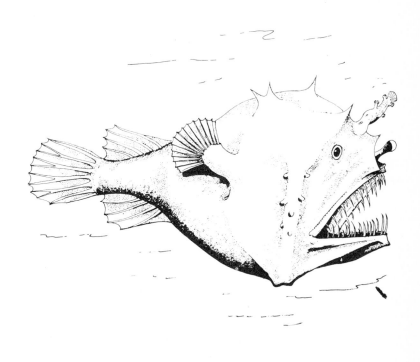

Female Deepsea anglerfish Photocoryne spiniceps *with a parasitic male fastened to her head. The males of such anglerfishes are much smaller than the females and spend nearly the whole of their lives attached inseparably to the females (see p80).*

A most unusual form of parasitism is practised by Deep-sea anglerfishes. One of the problems of the deep-sea environment is that the fishes are sparsely spread through a huge volume of water. Encounters between the sexes will not be very frequent so the most must be made of them. The Deep-sea anglerfish's solution is for the male to seize hold of the first female he meets. He grabs her with his teeth and eventually he fuses with her, being nourished from her bloodstream and degenerating to no more than a bag containing a pair of testes. He is, in effect, parasitic on the female, although this is not to her detriment. Joining of the sexes can take place while they are immature so no opportunity of gaining a mate is lost.

Filter-feeders The commonest method of bypassing the labour of chewing, one found especially in the lower animals, is to feed on small particles. This is known as filter-feeding. The particles are of small size, relative to the animal involved, and may consist of small plants and animals, or of bacteria, or may be what are called organic particles. The dead bodies of plants and animals decay, if they are not first eaten by scavengers, and in water they become broken down into smaller and smaller particles. Many aquatic animals, for example oysters, mussels and clams, are filter-feeders. They and others keep the water clean, as shown by the story about the Millport biological station.

This station is situated in the Firth of Clyde, in Scotland, and as is usual in marine stations, there is a line of tanks open at the top to greet visitors. One tank displayed Sea anemones, another fishes and a third contained only mussels. It seemed to have no water in it, and one visitor lowered his face to take a closer look only to find he had dipped it into the water. Because the mussels had extracted all particles from it, the water was as clear as gin.

The simplest filter-feeders are sponges. Some scientists have claimed that sponges are not animals. Certainly, their way of life is unusual. In life, the skin covering the body is perforated with numerous minute pores, with here and there a larger more obvious opening. The body of a sponge is traversed by a labyrinth of canals. At intervals are grouped special cells, each bearing a protoplasmic whip, surrounded by a protoplasmic collar. These are known as collared cells. These, by their lashing, create a current which draws water in through the pores and out through the larger openings. Food particles carried in by the stream are trapped in the collars and absorbed by the cells.

A well-known rotifer or wheel-animalcule, Melicerta *captures food with the cilia on its head-lobes and builds its tube with pellets of undigested remains.*

80

Another simple filterer is the mosquito larva, which lives in ponds and sluggish streams. Around its mouth are two small brushes of densely-packed bristles. These beat rapidly in a circular scooping motion that sets up a current of water towards the mouth. Bacteria and organic particles become trapped in the bristles, passed to the mouth and swallowed. The bristles form a simple strainer. In this way the mosquito larva swallows millions of bacteria and other particles without imbibing water except incidentally.

The animal filtering apparatus can, however, be very complex, as it is in oysters. An oyster breathes by gills. Each gill is a fine lattice-work and is covered by microscopic hairs of protoplasm known as cilia. The cilia beat rhythmically, in waves, looking as tall grass does when the wind passes across it. The moving wind makes the grass stems bend; the bending cilia make the water move and so set up a current. Water is drawn into the oyster shell and passes through the interstices of the gills. Particles are trapped in slime on the gills and passed to the crystalline style, which has been called the most remarkable structure in the animal kingdom.

The style is a rotating, jelly-like rod on which strings of slime laden with food are wound like a rope on a capstan. The rope is pressed against the wall of the stomach. Wandering cells pass through the wall of the stomach, seize the particles, pass back through the stomach wall and carry the food to all parts of the body.

This may sound complicated, but as written here, it is over-simplified. For one thing the crystalline style is kept rotating by the action of cilia lining the sac in which it is situated. Also, the outer end of the style is continually liquefying and the liquid that forms from it assists digestion in the stomach. It is being constantly renewed at the other end except when the oyster is closed. Then the style disappears altogether and has to be renewed as soon as the oyster opens its shell and begins to feed again.

There are many small aquatic animals known as wheel-animalcules or rotifers. The first of these two names could be misleading without some explanation. Rotifers are easily found in ponds and the microscopists of the nineteenth and early twentieth centuries took great delight in watching these animals under their microscopes. Around the mouth of each rotifer is a circlet of cilia moving all the time with a wave-like action so that it looks as though there is a wheel revolving. This causes a vortex which draws food down into the animal's mouth.

While so many of the filter-feeders use cilia in some form or other, there are other kinds of filter-feeders using different methods. The huge Basking shark, and the Whale shark which may be as much as fifty feet long, both feed on small plankton. Both have an enormous mouth which they open, letting the lower jaw drop, as they swim slowly through the water. The water rushes into the enormous mouth and passes out through the gills. On the gills are a number of semi-rigid rods known as gill rakers which criss-cross and form a sieve. This filters the plankton from the water before it reaches the gills. Something of the same kind is used by whalebone whales including the Blue whale, the hundred-foot leviathan which is almost certainly the largest animal that has ever lived. The whalebone whales have no teeth. Instead they have horny plates of whalebone suspended on either side of the mouth from the upper jaws. These are closely set together at right angles to the line of the mouth and on their inner edges they become frayed. The whale takes a great gulp of water. Then it closes its mouth and with its tongue drives the water through the whalebone out of the sides of the mouth so that the krill, small shrimp-like crustaceans, are left behind on the frayed edges.

Another kind of filter-feeder is the larva of the insect known as the Caddis fly. The adult can be seen flying near rivers, the larva lives in the water and is known as a Caddis worm. One feature of the Caddis worm is that it makes itself a tube of small pieces of leaf or tiny pebbles and lives inside this. The tube is reinforced with strands of silk, and some Caddis worms, instead of crawling laboriously over the bottom of the river dragging their little houses with them, spin a net of silk, fastening it to anything solid, at right angles to the flow of the river. For centuries fishermen have fastened nets across tidal rivers and collected the fish caught in the nets after the tide had gone down. The Caddis worm beat them to it by perfecting this method of obtaining food long before they did. The net of the Caddis worm is so like the web of a spider, into which insects fly and are caught, that there could be some justification in speaking of web-spinning spiders as filter-feeders.

One of the oddest filter-feeders is the Lattice-winged bat of Trinidad, so called because clear bands of skin in its wings produce a lattice-like effect.

The Lattice-winged bat has one of the ugliest faces of all bats but is harmless except to crops of fruit. It feeds on overripe and mushy parts of bananas and pawpaws, and

Feats of drinking

A very thirsty Arabian camel has been recorded as drinking 50 gallons of water at one time, taking just over ten minutes to do so.

A donkey has been known to drink 15 gallons of water in one bout. If we allow the volume of its body to be 60 gallons, this drink of 15 gallons is a quarter of the body volume.

A hedgehog, a much smaller animal, with a body volume of half a gallon, has been known to drink half a pint of water, or one eighth of its total volume, at a sitting. As the hedgehog walked away it staggered — and no wonder.

The human record is puny: a man has drunk 22 pints of beer — without quitting the room. (The accepted record is 31 pints of beer in 56 minutes by Philip Davies, aged twenty-nine, in Lancashire, England, in April, 1971. His body weight is not recorded.)

Tearful turtle

'Water, water, everywhere, nor any drop to drink' was the Ancient Mariner's comment on the hopelessness of being at sea without fresh water. Sea water is dangerous to drink because the high concentrations of salt upset the equilibrium of the body fluids which the kidneys cannot restore.

Seabirds, such as gulls and penguins, have overcome the problem by secreting a salty fluid from large glands above the eyes to reduce the strain on the kidneys. The fluid flows down the tear ducts into the beak. It can be seen sometimes collecting as a 'dew-drop' on the tip of the beak and is cleared by the bird shaking its head. Sea turtles, Sea snakes and perhaps the seals have a similar mechanism which is called **extrarenal** (literally 'outside the kidneys') **salt secretion**. Their tears run down the face.

Living without air

Oxygen is essential to life. It is needed to oxidize carbohydrates to carbon dioxide and water and to release energy for use in growth, muscle contraction and other body processes. But we can do without oxygen for a short time. During a sprint, an athlete barely breathes. He is respiring anaerobically, converting carbohydrate into lactic acid with release of energy for his running. At the end of the race, he pants, gulping in oxygen that is needed to oxidize the lactic acid down to carbon dioxide and water. He is repaying the **oxygen debt** that was incurred during the race. There is a limit to the size of the oxygen debt: the athlete's speed depends on his ability continuously to pump oxygen to the muscles and remove carbon dioxide from the body. Some animals can live with a permanent oxygen debt. Parasites, such as tapeworms in the intestine, live in an environment devoid of oxygen. They have to respire anaerobically and some insect larvae, such as Bot flies living in the tissues of mammals, survive by producing their own oxygen. They convert their carbohydrate (glycogen) store into fat and use the oxygen set free for aerobic respiration. Seals and whales dive for long periods (over 40 minutes for the Weddell seal and over 75 minutes for the Sperm whale) by conserving oxygen through shutting down inessential processes, such as digestion, and incurring huge oxygen debts. Desert lizards are sometimes flooded out by thunderstorms and they bury themselves in the sand; by lying still the Earless lizard of America survives for about a day without oxygen.

although it has teeth they are not very strong, perhaps sufficient only for opening the fruit. After this the bat begins to suck the soft, semi-liquid pulp between its lips. The lips and gums are covered with many small fleshy pimples, and when the mouth is only slightly opened these form a filter through which the juice is sucked.

Drinking

As well as eating, most animals except those living wholly in water, must drink. Eating and drinking are dissimilar processes although often associated even to the point of appearing complementary. Eating provides the fuel to drive the living motor. Food taken into the body is broken down, or digested, and its products used in building the body and in supplying energy. Drinking does none of these, it replaces lost fluids and is vital to many body processes. Water is being continually lost through the urine, the sweat and the moisture in the exhaled breath.

The amount of drinking required varies enormously in different animals. Some appear to be able to go indefinitely without drinking, obtaining water from their food and making the maximum use of this by water conservation devices. Others are able to store water when an abundant supply is available, while at the other extreme are those that must drink at regular, usually frequent, intervals. Blood-sucking and sap-imbibing animals have adequate supplies from their food and some, like the aphids, have the problem of getting rid of excess fluid. Aphids dispose of it in the continuous flow of honeydew. The human body can stand very little dehydration before things start to go wrong. Extreme thirst in man means that water is being drawn from the body and in particular from the blood, which becomes thicker and more turgid, throwing a strain on the heart.

Normally, for animals in temperate latitudes, drinking rarely poses a problem. But desert-living animals must conserve water. The first desert animal to be fully investigated was the Kangaroo rat of the deserts of the southwestern United States. This small rodent had been kept in captivity and fed only dry seeds, with no water to drink. It seemed to be able to go indefinitely without drinking.

Around 1950, K. and B. Schmidt-Nielsen investigated the Kangaroo rat and subsequent researchers established that their findings in this rodent set the pattern for many small desert animals. The first step in water conservation by the Kangaroo rat is to stay in a burrow by day and come out to

feed by night, so escaping the heat of the day. Secondly, the animal draws some water from that absorbed in the seeds it eats, but a larger quantity is obtained by metabolic oxidation of the food in the body. Thirdly, the Kangaroo rat has no sweat glands. Finally, it has several adaptations for preventing water loss. Its droppings are hard and dry, its urine almost solid and as it exhales the breath is cooled in the nasal passages and condensed water is retained to be passed back into the body.

All birds must drink and desert birds generally contrive to live near water except for sandgrouse. These make daily journeys of up to twenty miles to water and when they have young the males take water back to the chicks in their feathers. This was reported long ago and then disbelieved. It has now been substantiated, however, that the parents walk into water while drinking, so wetting their breast feathers, and the chicks strip the water from the feathers of the male parent on his return.

There are only scattered observations of other birds carrying water to their young. Several species of waxbills drink after filling their crops and then feed (and water) the young. Pigeons feed their squabs at first with 'pigeon's milk', a semi-liquid secretion from the lining of the crop.

Bats' meals

It is now well known that bats feeding on the wing detect their insect prey by echolocation. Even so they have only small mouths and the insects they take are often themselves very small and, one would have thought, difficult to catch. Electronic flash photography has shown that bats will take even the tiny fruit flies. They sometimes catch these directly with the mouth but usually the insects are dropped in the pouch formed by the membrane between the bat's hind legs and tail, the inter-femoral membrane. The bat then picks these out with its teeth, often doing so while in flight. Bats will also flick insects with their wings either towards the mouth or towards the tail pouch, after having located them by echolocation.

Some bats will land on the ground to feed, especially those that live on beetles. The Mouse-eared bat (*Myotis myotis*), of Europe, feeds largely in spring on running beetles, spiders and insect larvae. Investigation has shown that this bat can hop on the ground, can apparently hear insects moving on the ground, and is also able to locate them by smell. It will push its nose into moss to get the beetles. One experimenter tried to fool the Mouse-eared bat with a wad of cotton wool soaked in the juices of a squashed Dung beetle. This was hidden in the moss and the bat located it and started to chew it, but spat it out.

Bats usually drink on the wing, dipping to the water's surface; but some have been seen to land on the edge of a pool in order to drink or even to drop into the water and then paddle across with the wings before taking off again.

Some bats spend only two hours a day feeding, one

hour at sunset, and one hour just before dawn. In one feeding session such a bat may eat one quarter to one third of its own body weight. One naturalist who shot a Leisler's bat an hour after sunset 'found it so crammed with food that it did not appear physically possible for it to feed longer.'

Desert survival

Desert animals survive by their ability to last for long periods without water. They cut down their water losses by producing very little urine and by letting their body temperature rise without sweating; but the smaller animals in particular avoid the heat by burrowing during the day and feeding at night. It is surprising, however, to find that frogs and toads live in deserts. All amphibians exposed to dry air lose water very rapidly through the skin, but desert amphibians can lose up to 45% of their body water before they die. In practice, water loss is kept to a minimum because the amphibians burrow deep into the soil, where there is plenty of moisture; and they can survive extreme drought because they can store up to 50% of their body weight of water in their bladder. When the body begins to dry up, the losses are made good by water being absorbed from the bladder. The Holy cross or Catholic frog (see text) is sought as an emergency water supply by Australian aborigines in the desert.

Later they take water to them as well as food. Jays sometimes drink after filling the gullet with food for the young and ravens, in very hot weather, bring water in the gullet for the young. Probably more birds 'water' their young in one or other of these ways. During the very hot summer in 1975, in England, there was a positive observation of a blackbird taking water from a bird bath in a garden and dropping individual drops into the upturned beak of a fledgling. This had never been recorded before for that species; and blackbirds are among the commonest garden birds.

Reptiles, insects and spiders have a relatively impermeable covering to the body and are also capable of converting body waste to uric acid which requires little water, being highly insoluble, for its elimination. Frogs and toads usually live in damp places but there are desert species able to store water during a rainy period, then burrow into the sand to tide over a dry season. They have a large bladder for storing water. Also they shed the top layer of skin and this remains loosely wrapped around the body as a sort of cocoon. Water accumulates inside this. The best known is the Holy cross frog, or Catholic frog, of Australia, so named for the black cross on its back.

Several of the larger desert mammals, such as oryxes, eat succulent plants, which themselves store water in their tissues. The white oryx of Arabia has been known to go without drinking for ninety days. The camel cannot equal this but it is a specialist in water conservation. It does not store water in its hump, as is often said, but it can store water in the body tissues generally. It can lose over a quarter of its body weight in water, as compared with one-eighth in a man. Although it looks emaciated it suffers no distress because none of the lost water has been drawn from the blood. So its heart, brain and other essential tissues continue to function normally. Then, when water is available the camel can drink twenty-seven gallons in ten minutes, at the end of which it is no longer emaciated but bloated.

Chapter Four

Escape from Enemies

In the last chapter, the way in which an animal finds its food was the subject under discussion. In this chapter the problem is how an animal avoids becoming food. Few animals are such super-predators that they never have to worry about becoming the prey of another. In the final analysis, man is their enemy. As a general rule, the best method of defence in the animal world is to run away. Attack is reserved for occasions when the pursued animal is cornered and cannot escape or is defending its family.

There is a dreaded word in the vocabulary of the zoologist today. It is the word 'anthropomorphic', derived from two Greek words, *anthropos* and *morphe,* meaning 'man' and 'shape' respectively. Originally, it meant the habit of thinking of God in the shape of and with the attributes of a man. Today, it means attributing human emotions and motivations to animals. It is a dreaded word among some scientists because it has become almost an expression of derision; there is no more heinous sin in their eyes than to be anthropomorphic, as when we speak of an animal hating something, fearing something, having greed or showing lust. Nevertheless, it is very easy to drop into using such terms. It may be as well therefore to put this word into perspective.

Gazelles chased by a cheetah: though probably the fastest of land animals, it must make a capture in the first swift sprint; it has no staying power.

Flies and the swatter
In 1957, Mr. Don Ollis, professional photographer of Santa Barbara, California, set up his apparatus to investigate by what manoeuvres a fly manages to evade being swatted. By patience and ingenuity he obtained a series of photographs which tell the story. As the swat is descending, its movements are recorded on the eyes of the fly, which jumps backwards, then leaps into the air thrusting its wings downwards at the same time to assist the take-off. Then it banks and continues to fly upwards, helped by the blast of air as the swatter strikes the solid surface. Moral: strike behind the fly.

Even that may not be enough, for the fly may alter its tactics according to the movement of the swatter. If the disturbance comes from behind it, the fly will take off heading forwards and climb at an angle of 30-45 degrees. If the blow is aimed to the side of it, it will take off forwards, then immediately change course away from the swat.

Early warning systems
Barn owls have keen eyesight. Their eyes are at least thirty-five times as sensitive as ours and they can pick their way through woodlands even on moonless, overcast nights. But their hearing is also remarkably acute and Barn owls can pounce accurately on mice even in a completely lightless room, provided that the mice rustle through leaves or make some other sound. Only two rustles are needed to seal a mouse's fate. The first alerts the owl and the second shows it where to strike.

The mouse also has sharp hearing but the owl can swoop silently onto it because its feathers are covered with a velvety pile that damps the sound of the slipstream. Nevertheless, owls still make some sound and Kangaroo rats and other desert-living rodents have ears specially tuned to sound frequencies produced by a striking predator such as an owl or rattlesnake. Just before impact, the rodent jumps clear.

A similar defence system is used by some moths whose ears are particularly sensitive to sounds of 15-60 kcs, the range of the ultrasonic squeaks that bats use in echolocation. The moth's ears have only two sense cells apiece, yet they can detect the approach of a bat so that the moth can take evasive action, by jinking or dropping to the ground.

If a man runs away from danger, he is said to be afraid. To what extent, then, can the emotion of fear be applied to animals? Although it is usually considered anthropomorphic to ascribe emotions to animals, our emotions are our primitive urges. Hate, fear, greed and lust are sophisticated versions of aggression, the desire to escape, hunger and the drive to mate. Certainly, they share the same basic motivation; and the body's physiological responses are the same in man and animals. In the case of fear, the frightening circumstances prepare the body for flight, as was noted in Chapter 1. There is a sudden discharge of adrenalin into the bloodstream that puts the body on the alert. Blood drains from the face because it is needed in the muscles, the heartbeat rises, sweat begins to flow. Erector muscles around hair roots contract so we feel our hair stand on end — the equivalent of an animal's hackles rising. Perhaps the difference between man and animal is that we think about our fear and about what is frightening us, and this compounds our fear. An animal, on the other hand, reacts automatically and follows the actions which evolution has laid down as best for the species in such circumstances. When we say an animal is showing fear or panic it is because it is showing the same physiological symptoms that we recognize in ourselves and is abandoning its normal cautious behaviour.

Recognizing an enemy

The first step in escape from enemies is to recognize what is an enemy. As we saw in Chapter 1, an animal cannot react to every stimulus that might give cause for alarm. It has to habituate, or get used to those which it finds to be harmless. Knowing what is dangerous, before it is too late, is mainly instinctive but, in higher animals at least, learning also plays a part. A housefly instinctively takes wing when it senses the air disturbance presaging the arrival of a swat and young birds know the shape of a hawk flying overhead as soon as they hatch. The inherited nature of predator recognition is well shown by Darwin's finches, a family of birds which lives on the Galapagos Islands. They are descended from an ancestor that lived on the South American mainland and some fourteen species have evolved in the isolation of the archipelago. The mainland ancestor was subject to predation by many enemies; but the only natural enemies of Darwin's finches are a hawk, two owls and some snakes. The finches have retained a fear of these predators, even on Wenman

Island where they do not exist, but they have lost their instinctive fear of mammalian predators. This has been unfortunate as they have no fear of introduced domestic cats and, on Indefatigable Island, one species of ground-living finch has been exterminated by them.

Instinct and learning work together in predator recognition. American Song sparrows and European Song thrushes react instinctively to owls but if a model owl is presented repeatedly, their response wanes. In anthropomorphic terms, because the model does them no harm they soon grow tired of being fooled. If this were not so the growers of crops, especially fruit crops, would have no problem. As it is, the use of scarecrows and other such devices fails to be fully effective, so that it is not unusual to see birds perching on scarecrows in the intervals between attacking our crops. However, in natural circumstances they will not see owls often and habituation will not take place. Song sparrows have to learn fear of cats and snakes and European Song thrushes have to learn fear of man. Newly fledged birds are fearless but they soon learn from the alarm calls and escape behaviour of their parents. Why man should be so universally feared is not known for certain. It may be that, unlike other animals, his behaviour does not follow stereotyped patterns. His appearance is unexpected, like a hawk's, rather than regular like a quietly browsing deer. Animals on remote islands, however, frequently show no fear of man and, after ample opportunity to learn, still remain tame. This problem will come nearer solution when we have a better idea of what **signals** elicit fear. However, as a general rule it can be said that animals, and man, fear anything new. Whether they continue to fear that thing depends on what they learn about it. Wild animals lose their fear of humans if they are allowed to become used to them, in the way that the wild chimpanzees learned to accept the presence of Jane Goodall when, for study purposes, she lived among them over a period of many months.

Signals Signals vary from one species to another. The Song sparrow is instinctively frightened of owls but not of snakes whereas the Curve-billed thrasher has an inborn fear of snakes and has to learn about other predators. Response to predators also varies because the same predators may give different signals at different times. Grazing antelope do not flee from lions that are in sight and merely look alert if a lion walks by. They may even approach it. But they react when a lion is on the prowl in the bush and always avoid entering the

Recognizing the threat
As with the Black-headed gull (p 91), the reactions of small birds depend on the threat posed by a predator. Pied flycatchers give a snarling call when disturbed by predators, such as squirrels and woodpeckers, that are a danger to eggs and nestlings, but they mob (or harass) predators of adult flycatchers. They gather around a perched owl or sparrowhawk, calling and flicking their wings and tails. There is no attempt at concealment and the purpose of mobbing seems to be to draw attention to the position of the predator and to cause young birds to remain still and silent. Mobbing is most common during the breeding season and some small birds give it up for the rest of the year.

A hawk in flight elicits a different reaction. Pied flycatchers 'freeze' to avoid alerting the hawk if it is far away but if it is near them, they flee to cover. The calls given by small birds in danger from flying predators differ from that elicited by perched predators. They have acoustic characteristics that make them very difficult to locate. The narrow frequency spread of the 'cheer' call of American Red-winged blackbirds, is similar to the hawk alarm calls of European blackbirds, tits and chaffinches.

It is often said that the similarity of cuckoos to sparrowhawks helps the cuckoo in its ploy of laying its eggs in the nests of small birds. But Willow warblers recognize a cuckoo as a threat to the brood and the hawk as a threat to adults and react with the appropriate alarm calls.

Instinctive fear of hawks

The chicks of ground-nesting birds leave the nest at an early age. Young ducks and pheasants then lead a nomadic life, wandering about in search of food under the watchful eye of their mothers. At the slightest disturbance, the chicks either crouch where they are or run for cover. One thing that alarms them is the silhouette of a hawk with its short neck and long tail passing overhead. At first anything in the air causes alarm, even a falling leaf, but the chicks soon habituate to common objects. They learn that the common kinds of birds are harmless but hawks are seen so infrequently that there is little chance of habituation taking place. There is also evidence that they are born with the ability to recognize a hawk. If models are flown over them, newly-hatched ducklings are alarmed by anything that flies at the same speed as a hawk. Also, a flat, cut-out model that looks like a hawk when flown one way but looks like a long-necked, short-tailed goose when flown the other way, elicits much more alarm when hawk-like.

The domestic hen gives a special alarm call when a hawk flies overhead, or even when its shadow crosses the ground in front of her. The chicks then seek cover under her fluffed out feathers. The hen will also give this hawk alarm call when an airplane passes overhead — provided it does not have a long nose to the fuselage.

bush themselves. Then again, the reaction of the preyed-upon animal may depend on the sort of threat that different predators pose. Black-headed gulls flee from falcons that prey on adult gulls but fearlessly attack crows and hedgehogs which eat only their chicks and eggs and cannot harm the adults. Foxes and men pose a dual threat. They kill adult gulls and also rob their nests. Accordingly, the gull's reaction is **ambivalent** and they treat these predators with a mixture of aggression and fear which results in the well-known **stooping.** If a man or a fox enters a gull colony, the birds swoop down at them but shear away at the last second.

A predatory animal can approach to within a certain distance of its prey before the latter takes flight. This distance between the two is known as the **flight distance** and varies considerably. Some animals are 'tamer' than others and circumstances control flight distance, as in the case of the antelopes and the well-fed lions. An even better example can be seen in the Wood pigeon, which is a native of the open country. It has, in places, invaded large towns and their parks in small numbers. To take London as an example, in the rural areas surrounding it, where the Wood pigeon is leading a 'natural' life, the birds are likely to take wing at the approach of a man when he is perhaps fifty yards away.

91

In the suburbs, where conditions can be described as semi-rural, one may approach to within twenty yards before the birds take wing. In the centre of London, in the streets with heavy traffic, Wood pigeons may sometimes be seen feeding unconcernedly as people pass within a few yards of them. They have become habituated to the presence of people.

Many naturalists make use of the fact that animals are tolerant of humans if they stay in a car and in Africa the Land Rover is widely used as a convenient mobile hide. Jane Goodall found that her chimpanzees originally ran off when she was five hundred yards away but after fourteen months she could sometimes get within ten yards. Approach was even easier if she disguised her interest in them by pretending to be absorbed in something else.

Methods of defence

Defending the family against an enemy is **altruistic behaviour.** The animal is putting its own safety in jeopardy in order to defend others. To our eyes this is heroism but the animal is usually careful not to go too far. Biologically, it is better to fight and run away and live to breed another day, rather than to make the supreme sacrifice, because then the orphans will also succumb. There are, however, some stories of mothers dying to save their offspring. Herbert Ponting describes such an incident in *The Great White South*. He was filming from the deck of the *Terra Nova* as she was moored to the ice. A school of Killer whales approached and a Weddell seal sprang onto the ice, turned and peered into the water. Suddenly, her pup appeared alongside but it was unable to pull itself out of the water. By this time the killers were only a dozen yards away and, with no more ado, the mother seal slid into the water and lured the whales away. The pup was still struggling helplessly when the whales returned and, again, its mother led the Killer whales away. This time she returned and lifted the pup onto her back but, just as it reached safety, it rolled off into the sea. It was now too late to do anything and both seals disappeared as the Killer whales finally closed in on them.

A communal form of altruistic behaviour is sometimes exhibited by animals living in open country where escape is difficult. Muskoxen of the Arctic tundra band together in a protective phalanx or *Karre* when threatened by wolves. The bulls form a close wall facing the enemy with the cows and calves huddled behind them. At intervals a bull will emerge to harry the wolves and then return to its place.

Flashing with fright
Fireflies use a living light to signal to each other. Males use a different sequence of flashes to signal to females so the flashing lights can be regarded as a sort of court-ship, or at least a means of bringing the sexes together.

The production of living light is something of a mystery in two respects. First there is the production of light without heat, the secret of which for so long baffled the world's scient-ists. Secondly, it is not al-ways clear why a particular animal is using this biolumi-nescence.

Two Harvard zoologists have gone some way to probing the secrets of a small shrimp, *Metridia lucens,* which is $\frac{1}{8}$ in long. When touched or disturbed this mitey shrimp gives out a flash of light which rises rapidly to a peak of brilliance and then fades away. The light is produced from several glands on the front of the head, on the back and on the tail-end. When some of these shrimps were put into a tank containing larger shrimp-eating shrimps they gave a brilliant display of their light-producing quali-ties, and although a few were eaten by the larger shrimps it was clear from the number of flashes they all produced that many had escaped capture as a con-sequence of giving out light. Probably a bright flash produces an after image in the eye of a shrimp just as it does in our eyes and this would momentarily baffle them and increase the chance of their small prey escaping.

Females of the firefly genus *Photuris* feed on other in-sects including other fire-flies. They have a trick of mimicking the mating flash pattern of the females of other species so luring the males to them, to be de-voured.

The killdeer of North America, like other plovers, feigns injury to draw a predator away from its eggs or chicks.

Evil Eye

In the early years of this century a naturalist experienced in the way of the Indian jungle was astonished to see monkeys falling out of a tree and lying in a dead faint, as he put it, on the ground. One monkey would come hurtling down from the tree, then another. The naturalist stood and watched, hoping to find the cause of this extraordinary behaviour. Then his glance happened to alight on a part of the undergrowth a little way from the base of the tree. In it he could see the yellow eyes of a leopard gazing fixedly up at the monkeys.

A similar episode was reported some years later from Africa and in more recent years monkeys were seen falling from a tree in South America while a jaguar, hidden in the undergrowth, gazed up at them.

Unfortunately, this form of defence is suicidal when employed against hunters carrying firearms. Yellow baboons of the East African plains have a rather similar formation. They travel with the old males, females and young in the centre of the band surrounded by the young males. Their enemies are lions and leopards and, when one is sighted, a shrill bark of alarm is given. The largest males run over to keep watch and even attack.

Ground-nesting birds have evolved special **distraction displays** to lure predators from their nests by putting their own lives at risk. In the simplest form of this, the bird, such as a pipit, flutters on the ground in front of the predator. The movement attracts the predator who attempts to catch the bird, which escapes by flying up. Repeating the sequence leads the predator well away from the nest. A more elaborate display has been called **injury feigning** or the **broken-wing display**. The bird flaps one or both wings against the ground or runs in front of the predator and makes itself more conspicuous by drooping one wing to reveal the white plumage of its flank. Ducks and waders (shore-birds) are well known for this behaviour and, although it looks as if the bird is injured, it is doubtful whether the predator sees it this way. The value of the display lies in its conspicuousness, due to the movement and the show of white plumage, which will attract the predator.

Also altruistic but carrying less danger is the giving of **alarm calls** to alert other animals to the danger. Alarm calls are used particularly by animals that live in flocks or herds. Living together gives them a measure of safety because many pairs of eyes keep a better watch than one, but the detection of danger must be communicated to the others if the increased watch is to be effective. In some instances, particular sentinels keep watch while their comrades feed. Prairiedog sentinels sit back on their hind legs to get a better view over grassland. A sharp bark, from which the prairiedogs get their name, sends the other members of the colony scuttling to the safety of their burrows. In European gardens and woodlands, the thin 'seet' call of a chaffinch is the signal that there is a hawk overhead. All birds within earshot dive for cover. The 'seet' call is a sound of uniform pitch which gradually waxes and wanes. Both these properties make the call very difficult to locate and it is no coincidence that a variety of songbirds, such as blackbirds, buntings, tits and finches, have similar calls. This contrasts not only with the birds' territorial songs which are designed

A herd of topi, East African Bastard hartebeests, with one of their number standing on a 'hill', acting as sentry.

to advertise the position of the singer, but also with the scolding, 'chink' alarm calls which are given when cats and other ground predators are prowling nearby. 'Chink' calls can be located, so they give away the position of the intruder.

Distress calls are given by an animal when it has been captured or wounded. They have the same warning function. Taped distress calls of gulls have been broadcast at airports in an effort to drive gulls from the runways. Many animals remain silent when in distress but hedgehogs emit an unearthly scream when trapped and the calls of young animals, in particular, attract their parent's attention. Among adults, too, distress calls may attract comrades rather than send them fleeing. This seems to be particularly developed in dolphins.

Distress signals also come in the form of pheromones. When a minnow is seized by a pike, the ruptured flesh liberates an **alarm substance** which causes nearby minnows to flee. Until the alarm substance disperses, the minnows continue to flee at the slightest disturbance. Related fishes react in different ways to alarm substances. Bottom-living tench, for instance, plough through the mud to work up what is in effect a smokescreen.

Shy larks
Larks, with the exception of the Black lark, are brown streaked with grey and buff, with paler underparts. They are not easy to see when standing motionless on the ground and in some desert regions, where concealing vegetation is scarce, larks are often very well matched with the soil on which they live. A survey of twenty-two Thekla and Crested larks showed that twenty had plumage that matched the soil where they were nesting. This suggests that the larks carefully select their background. In Arabia, Desert larks range between pale and black over the space of a few miles where there are patches of dark lava set in the pale desert sand. If a Desert lark is chased it will refuse to cross onto the wrong background.

Click beetles
A sudden movement on the part of a prospective meal is often sufficient to startle a predator into missing its mark. Surprise together with rapid flight is combined in a neat mechanism by the Click beetles or skipjacks. These common beetles, whose larvae are the notorious wireworms that feed on plant roots, have short legs. They find it difficult to rescue themselves from a supine position by the normal insect method of reaching out with the legs and, when anchored, hitching themselves over. Instead, the Click beetles fling themselves into the air with an even chance of landing the right way up.
To jump, a Click beetle arches its back against the ground by bending the joint between the first two segments of the thorax. A spine on the first segment is brought out of a groove in the second and rests on its lip. All the tension in the arched body is now taken by the spine which is held in place by friction. When released, the spine snaps back into its groove with such force that the beetle is thrown into the air. From a hard surface, leaps of a foot have been recorded.
When disturbed, a Click beetle drops to the ground. There, it arches itself and further molestation causes it to leap into the air, surprising an enemy and perhaps being carried clear of it. If a beetle is picked up it repeatedly 'clicks', which may be enough to startle a predator into dropping it.

Alarm substances are used by ants in such a way that it looks as if their behaviour on being disturbed is purposive. If a nest of Wood ants is disturbed, the workers squirt formic acid from the tip of the abdomen. The acid evaporates and diffuses through the air to alert other ants within a radius of a few inches. These ants run about with their jaws open as if looking for an intruder and they also squirt formic acid. If the disturbance disappears, the ants stop spraying, the formic acid disperses and peace returns to the nest. If the disturbance continues, more ants are recruited to its defence and the air soon reeks of formic acid. Therefore, the alarm system is organized to gear the size of the defence to the magnitude of the disturbance. The Harvester ants of Florida have a more elaborate system. They are attracted by low concentrations of pheromones so they gather at the disturbance. As the concentration increases, they start to defend the nest with vigour until a very high concentration, indicating a violent attack on the nest, causes them to change their behaviour and seek to hide.

Camouflage   Running from the enemy is by no means the only way of avoiding being eaten. The issue can be dodged by failing to attract the enemy's attention in the first place through various forms of camouflage or **cryptic coloration.** The first essential of camouflage is to keep still or to **freeze.** This is particularly valuable when the predator is still some distance away. As the predator approaches flight replaces freezing. Observations of animals as varied as mealworms, lizards and ducks show that freezing helps to protect them from being eaten.

The eyes of insects and most vertebrates are particularly well designed for detecting movement and we have seen that frogs in fact fail to react to stationary prey. This is important to both prey and predator. A lion stalking an antelope freezes every time the antelope raises its head and the chicks of gulls freeze when they hear an alarm call from their parents. By remaining still even a gaudily plumaged bird tends to escape detection while the brown and black patchwork of a gull chick blends beautifully with coarse vegetation. At first the gull chick crouches where it is but when a few days old it will run to a particular hiding place, among boulders or under a bush. If picked up it stays motionless and does not struggle. Chicks of the Greater black-backed gull are known to adapt themselves to their surroundings. Those that live among vegetation freeze but those living on stony ground keep running.

Some animals can change colour to match their background. The ability to change colour may be of vital importance to an animal in saving it from enemies. The change is due to movements of pigment cells in the skin, governed by nervous or hormonal control. It may be rapid or it may take days, but the end is the same, to make the animal resemble in colour the background against which it is living. The camouflage effect is often increased by the animal's shape, as in the chrysalis of the Passion-vine butterfly which looks like a withered leaf, or the familiar stick-insect that folds its legs against its elongated body and, holding itself rigid, is indistinguishable from a twig. Many fishes, such as turbot and plaice, which live on the bottom of the sea, have considerable ability to adjust their colouring to suit the background on which they are lying, usually mud, sand or gravel. They become vulnerable when they move to a new area, of different colour, where they are conspicuous until a new **colour change** has been completed.

To demonstrate the effectiveness of camouflage, experiments have been made in which a predator is let loose on a number of prey animals. The result is that the predator catches those that do not match their background. A matching colour pattern is only part of the story because the animal's behaviour is also important. For instance, if Water boatmen are placed in a tank with a sandy bottom, a minnow will soon pick out those that are easiest to see. However, this is not a natural situation because Water boatmen gradually take up the colour of their background and, if given the opportunity, they come to rest on the background which they match. Batfishes and leather-jackets of American waters not only mimic the appearance of yellowing leaves of mangroves, they also behave like waterlogged leaves floating in the sea. The fishes rest near the surface, hanging head down and twisted, and allow themselves to drift to and fro.

Facing the enemy

If the animal has to face the enemy instead of hiding or running away, it can resort to a number of ruses. It can threaten its enemy, by adopting menacing postures that indicate that it will defend itself or it may merely **bluff** the enemy that it is dangerous. Bluff very often consists of an animal making itself appear larger than life. The Frilled lizard of northern Australia and New Guinea has a large flap of skin that normally lies folded like a cape over its shoulders. Whenever possible the Frilled lizard escapes by

The toad's dilemma
The caterpillar of the Elephant hawkmoth of Europe leaves its food plant in the autumn in search of a place to pupate. It is three inches long, brown with four large eyespots behind the head. If disturbed in its quest, the caterpillar draws in its head and freezes. As the head retracts, the body segments behind it bulge and the eyespots are thrown into prominence making the caterpillar look like a miniature snake. If a toad has caused the disturbance, the effect of the caterpillar's reaction is dramatic. The toad blows up its body until it is half as big again. At the same time it rises on stiff legs.

As the toad is now stationary, the caterpillar slowly relaxes and continues on its way. It now looks like food to the toad, which also relaxes and moves into position to strike. The slight movement again alarms the caterpillar which retracts its head showing off its eyespots. Once again the toad reacts. And so it goes on until the caterpillar escapes.

96

The 'defenceless' Frilled lizard, of Australia, successfully uses sheer bluff to repel attackers, the main feature of which is the spreading of an enormous neck frill.

Losing a leg

The apparent ease with which a lizard or a crab can lose a tail or leg to escape the grip of an enemy (p 100) has long fascinated zoologists. The organ has a 'plane of cleavage', a line of weakness which readily ruptures when adjoining muscles contract, but, in crabs at least, it has been difficult to see how the animal avoids losing its legs during the normal muscular contractions of walking.

Autotomy is prevented by a 'safety clip' mechanism. During normal activities the tendon of the muscle that raises the base of the leg pulls on a strengthened flange of the cuticle forming the hard external 'skeleton'. When it becomes necessary to shed the leg, a smaller muscle that connects with this tendon contracts and tears it from the flange. The autotomizing system is now primed. The tension in the tendon is now directed on the 'clip', a plug of cuticle that fits in a socket across the plane of cleavage. When the main muscle contracts, the 'clip' is withdrawn. All that now holds the leg together is the weakened cuticle over the plane of cleavage, and this is snapped with a slight pull or twist.

Fatal bleeding is prevented by a membrane that runs across the plane of cleavage and seals off the blood vessel and nerve when these are torn apart.

running but, if brought to bay, it turns towards its pursuer, opens its mouth and spreads its frill by means of cartilaginous rods which act like the ribs of an umbrella. The lizard then looks like a huge, menacing animal and it will hiss and advance on its attacker. It is all bluff because the lizard has no weapons, yet dogs and presumably other carnivores are driven away.

Chimpanzees will use sticks or small branches to threaten an intruder. They may even use them to strike physical blows. More usually these are used as part of a bluff. A large chimpanzee angrily jumping up and down, with its hair standing on end and stick in hand, is a terrifying sight. These apes, in quarrels between themselves, also use stamping and **branch shaking** as threats. Gorillas, more powerful than chimpanzees yet among the most pacific of animals, use **chest beating** as a bluff.

Another kind of bluff is to **mimic** a dangerous animal. Elephant hawkmoth caterpillars, of Europe, bear pairs of large, round 'eye-spots' on their bodies. These are not very conspicuous until the caterpillar is disturbed, when it draws in the front part of its body and the false 'eyes' stand out, making the caterpillar look like the head of a snake, a terrifying sight for a creature that is expecting an easy meal.

Surprise is a vital element in these displays. A number of butterflies and moths, such as the Peacock butterfly, have large coloured spots on their wings. When disturbed, they spread their wings to take flight and so display their eye-spots. This is known as a **flash display** and to show that it is effective as a bird scarer, songbirds were trained to pick mealworms from a trough. Then small screens were placed on each side of the trough so that a variety of shapes could be flashed at an approaching bird by switching on a light underneath. The birds were not particularly bothered by the sudden appearance of pairs of crosses or bars but they were scared by circles suddenly appearing alongside the mealworm. They were really frightened if the circles were made to look like eyes by putting smaller circles, slightly off-centre, within the larger circles. This pattern, which makes the eyes appear to stare, is seen in the eye-spots of butterflies and moths. The importance of eyes in frightening a bird is shown by experiments in which chickens were presented with a stuffed hawk. If the hawk's eyes were covered with tape, there was no reaction but the chickens were scared by the presentation of a pair of detached glass eyes. An extra refinement of the Peacock butterfly is that it makes ultrasonic clicks as it flicks its wings, which scares bats.

The greatest bluff of all is to feign death because many predators do not eat carrion (that is, the flesh of animals already dead) or they fail to attack motionless prey. The expression **playing possum** comes from the behaviour of the Virginia opossum. If seized, the opossum suddenly lies limp with eyes shut and tongue lolling. It looks as if it has gone into a trance but electroencephalogram recordings of the brain's activity show that the opossum is still awake and very alert. **Feigning death** is quite a common habit. Many insects such as beetles and caterpillars feign death when knocked from the plant where they were feeding. They fall to the ground and lie quite unresponsive to touch, with their legs drawn up. Foxes, African Ground squirrels and hyaenas have also been recorded as feigning death. A hyaena feigns death while it is being examined but, as soon as the predator relaxes, the hyaena leaps up and dashes away.

Two snakes are particularly practised at feigning death. The American Hog-nosed snake and the European Grass snake are harmless and at first try to defend themselves by bluff. The Hog-nosed snake blows up its body and hisses with its mouth open wide. The Grass snake first emits an evil smell and then blows itself up and strikes with its mouth as if

Suspended animation in a bantam chick

Two bantam hens were sharing a nest containing 24 eggs in a chicken run beside a kitchen garden. One day one of the gardeners went down the garden at nine in the morning. He noticed three chicks in the run, so the eggs had started to hatch. Unfortunately, two chicks were obviously dead because of severe lacerations on head and body. The third was lying apparently dead — that is, when the gardener picked it up it felt cold and limp, so he threw all three on to the compost heap.

This was evidently the work of rats, since there was a rats' hole in the floor of the run, and furthermore there were now only 16 eggs under the two hens, so five had vanished. The killing must have been recent; probably the arrival of the gardener had disturbed the rats, and these had bolted instead of carrying their prey away.

The precaution was taken of removing the hens and putting each in an 'ark' specially used for hens with broods, and the eggs were divided between them. The arks were placed just outside the run, beside the compost box.

At 12.15 p.m. a second gardener took up an armful of weeds, walked over to the compost box, threw them on the heap, and was surprised to hear the muffled cheeping of a young bird. He thrust his hands into the weeds to part them and saw the bantam chick alive, vigorous and calling 'fit to burst'.

The chick survived to grow up and gave no sign of being any the worse for the adventure. There is every reason to believe that it had merely suffered from a state of suspended animation or, to use the more scientific word, **akinesis**, due to shock.

it were a venomous snake. If these ruses fail, both species roll onto the back in a contorted posture, with mouth agape and tongue hanging out. The snake looks utterly lifeless, and determinedly so because, if placed the right way up, it rolls onto its back again. The nervous mechanism of feigning death is discussed under hypnosis in Chapter 8.

Erratic behaviour Bluffing the enemy consists of presenting it with a different set of stimuli, so that it is faced with the message 'danger' instead of 'food' and its behaviour is switched from the appetitive behaviour of feeding to escape reactions, or, at least, it remains undecided long enough for its prey to escape. While not always being able to switch the enemy's behaviour, prey animals can often confuse it and upset its appetitive, or hunting, behaviour. Erratic behaviour prevents the predator from guiding its attack. The jinking run of a hare and the dodging flight of a snipe are examples. **Jinking** has another advantage in that a small animal is more manoeuvrable than the larger predator. As the latter catches up, the former can do a quick turn and gain a short lead. An exponent of this is the African springhare, a rodent that looks and moves like a miniature kangaroo. Its long hindlegs and bipedal gait allow it to change direction very suddenly.

Springbok pronking: this South African antelope is named for its habit of pronking when alarmed, leaping 8-10 ft into the air in a series of stiff-legged jumps before taking refuge in flight.

The confusing effect of jinking is increased when a group of animals perform the evolution en masse. On the appearance of a predator, gazelles make no attempt to flee but engage in **stotting** or **pronking,** which are alternative names for the same behaviour. The most spectacular displays of this are made by the aptly-named springbok. In stotting or pronking, the springbok leaps ten to twelve feet into the air with legs stiff and body curved. At the same time it erects a tuft of white hair on the back which acts as warning to other springboks who, by pronking themselves, further confuse the predator. It is too late for concealment so the gazelles make themselves conspicuous; the weaving of the white patches as the gazelles pronk to and fro and finally run off, leaping and jinking, is very confusing. To make a successful attack the predator has to select a victim and it is difficult to concentrate on a kaleidoscope of animals.

Exposing conspicuous flashes is a common weapon of defence among gregarious animals. The white rumps of deer and antelope, the white tails or 'scuts' of rabbits and the flashing scales of fishes are examples. Extra safety is gained by bunching together. A tightly packed flock of starlings or Black-headed gulls jinks as one body when threatened by a Peregrine falcon. Not only does the mass confuse the bird of prey, but it is put off by the danger of accidental collision.

A final method of side-stepping an attack is to deflect it by sacrificing an inessential part of the body. Although birds are scared by large eye-spots on the wings of an insect they are drawn to peck at small spots. Therefore we find that many butterflies have a row of small spots around the margins of the wings. These act as targets for a bird's pecks and it is not uncommon to find butterflies in which one or more spots have been pecked out. Lizards take this behaviour to the extreme by practising **autotomy.** There is a weak plane of cleavage across one of the vertebrae in the tail and, when muscles on each side contract, the tail parts company with the body. It wriggles violently for some time, so attracting the predator's attention while the lizard slips away or 'freezes'. The tail is gradually regrown but the new one never attains the size of the original. Crabs and other crustaceans also practise autotomy (see p 97).

Combat

If escape or any form of bluff fails, an animal may be forced to defend itself. Hoofs, horns, teeth, beaks, claws and stings are brought into effect. Even the slow-moving sloth can give

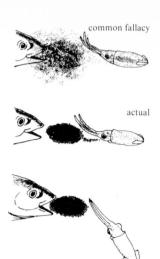

common fallacy

actual

Smoke screen fable
A zoologist on a Fisheries Research Vessel off Singapore was looking into a tub of seawater one evening in 1955. The tub was the usual container used on such ships for emptying the nets of their contained animals after being hauled through water. He saw a small squid in the tub and tried to catch it in his hand. As his hand approached the squid moved about, to escape, at the same time changing colour rapidly. When the zoologist's hand was within 9 inches of it, the squid stayed still and went black. The zoologist grabbed, but all he got was a handful of ink. So it was discovered, for the first time, that the age-long story of squid and octopus discharging ink into the water and retreating under cover of this 'smoke screen' was incorrect.
What had happened was that the squid squirted its ink and this remained as a dark blob about the size and shape of the animal itself. Meanwhile, the squid went pale, almost transparent, shot rapidly across the tub and remained still again, and was at first overlooked by its pursuer. A natural enemy would have been more baffled since it

would have had the ink between itself and its intended victim, and this would have slowly diffused through the water making vision even more hampered.

That ink from a squid or octopus does not immediately form a dark cloud, as had been believed for so long, can be tested by anyone who cares to take an eye-dropper, fill it with ink or any dark fluid and squirt this into a bowl of water. The ink stays as a blob for a few seconds and only then spreads out.

Lion-hearted mouse

One does not need to be a sentimentalist to be distressed at the sight of a cat playing with a mouse. To the human mind there is something particularly cruel and callous in the cat's actions and especially when, as sometimes happens, the mouse on being released from the cat's jaws turns round, squats on its haunches, turns its face up to the cat's face and brings its forepaws together as if in prayer. It looks as if the mouse is piteously begging for mercy, until we find that this is the aggressive attitude of the House mouse. The diminutive rodent is actually showing fight to its giant predator.

a good account of itself with its hooked claws. Each animal has a standard pattern of attack designed to make best use of its weapons. If a Sea otter, for instance, is prevented from reaching the safety of the sea, it faces the intruder then flops onto its back. If the intruder comes closer it tries to bite and push with its forepaws. Rolling onto the back is an unusual defensive reaction but it is probably related to the Sea otter's habit of lying on its back in the water. When afloat it can escape by diving but on land it has to defend itself.

Teeth, claws and beaks are characteristically used, defensively as well as offensively, by predators. Hoofs and horns tend to be used intraspecifically, that is, in contests between members of the same species, and although hoofs and horns are carried typically by herbivores that are prey species, both are at times used with deadly effect against predators. Zebra stallions sometimes kick out at lions and lionesses and may smash their teeth. The big cats are then forced ignominiously to eat small prey such as rodents, even insects, or anything alive and moving that they can catch and readily overpower. The hoofs of an adult giraffe are normally a match for teeth and claws, the offensive weapons of the lion. Oryxes, medium-sized to large antelopes, can use their sabre-like horns to impale or toss even the larger carnivores. One of the larger of them, the gemsbok, of southwest Africa, has been seen to lower its head and, with its long, gracefully curved horns, scoop up a lion and throw it over its back. The African buffalo excels in the use of horns and hoofs as offensive weapons. A herd of buffalo will trample a lion to death and there is a record of a herd passing a lion on their horns, from one to the other. When the lion was finally allowed to drop to the ground it was dead.

Unusual weapons Apart from the obvious weapons, there are some unusual means of defence in the animal world. Birds of the petrel family form an oily fluid in the stomach, consisting of partly digested food, which they feed to their chicks. However, petrels nesting on the surface rather than in burrows, such as fulmars, Giant petrels and Cape pigeons spit this evil, sticky liquid at intruders. As anyone who has attempted to handle these birds knows, it is a most effective deterrent but it is not in the same class as the liquid ejected by a skunk. The skunks belong to the family Mustelidae, along with weasels, otters and badgers. These animals discharge the contents of a pair of glands at the base of the tail when excited or disturbed but in skunks the glands have developed into truly offensive weapons. The Common

101

striped skunk can squirt its fluid accurately as far as twelve feet and can cause temporary blindness and choking.

Before squirting its fluid, the striped skunk warns its attacker by erecting its tail, lowering its head and stamping. The Spotted skunk does a handstand. Both postures display the striking black and white fur. It is a common feature of poisonous or distasteful creatures to have conspicuous colours or markings to act as a warning. Hence, there are the black and yellow stripes of wasps and Cinnabar moth caterpillars and the black and red of ladybirds and Burnet moths. An exception to this rule is the Australian stonefish, the most poisonous of all fishes, which is so very well camouflaged that it is almost impossible to see among the seaweeds and other growths on the seabed.

Discretion better than valour

When all is said and done, the most sure protection, for those capable of doing so, is to run away. This tactic is too familiar and commonplace to need examples to illustrate it. There is, however, one outstanding instance of a variety of escape reactions of this kind used in combination: this is the basilisk, a lizard living in Central America and ranging as far north as central Mexico. There are several species. Their vernacular name as well as their scientific name, *Basiliscus,* is based on the crest on their heads recalling that worn by the fabled basilisk, or cockatrice, of Europe. This legendary animal, which was reputed to kill with a glance, was portrayed as half cockerel, half snake.

The modern basilisk has no lethal powers, being timid and only too ready to flee. It lives along river banks, on bushes overhanging water. At first sign of danger it drops to the water and scutters over the surface, being assisted in this by a fringe of scales along the sides of its lengthy toes on the hind feet, whence the name Jesus Christo lizard from its ability to 'walk on the water'. It is also known as tetetereche, from the sound it makes in doing this. As its speed slackens the basilisk sinks lower in the water and finally swims.

Before discharging their malodorous secretions, skunks adopt a threatening attitude, which gives prior warning. In the Spotted skunk this takes the form of a handstand.

The basilisk lizard of tropical America scuttering across the surface of a river. It is named for this habit 'paso-rios' (river-crosser) or Jesus Christo lizard.

Chapter Five
Holding a Territory

Nearly all animals have a place to live — a home, if you like. They do not wander at will and the expression 'as free as a bird' is misleading. Each animal normally spends its life in a circumscribed area where it feeds, sleeps and rears its offspring. The form of the living space varies throughout the animal kingdom and, for each species, is intimately related to its way of life. The idea that an animal spends its life in one place was familiar to the Ancients. Aristotle was aware that eagles hunted within a particular region, but the study of living space started in earnest when Eliot Howard, a British ornithologist, published his findings on the territorial life of birds in 1920.

Definition of territory

A **territory** can be simply defined as 'a defended area' but we shall see that the territories of animals vary greatly in both form and function. By maintaining the integrity of its territorial boundaries against neighbours, an animal, or a family or clan of animals, preserves space in which to live. Territories are found widely in the animal kingdom; among many mammals, most birds, reptiles, a few

Decisive song contests

Howler monkeys, of South America, are like children, noisy first thing in the morning and noisy again before bedtime. Even the noisiest child cannot, however, vie with a Howler monkey for sheer output of sound. The monkey has the advantage of an expanded lower jaw enclosing a swollen throat within which the enlarged hyoid or tongue bones are accommodated. The hyoid bones form a resonance box and the voices of a howler troop raised in chorus at dusk and dawn more especially can be heard three miles away.

It seems these two sessions of chorus are not indulged in merely to assure their owners that their voice-boxes are in working condition. The chorus functions to let other troops of howlers know that the territory is occupied. There is also, in this, an implied process of agreeing the boundaries of contiguous howler territories (see p 114).

There is another way in which the voice is put to legitimate use by Howler monkeys. Should one troop come too near the boundary of the territory, or possibly overstep the mark, the occupying troop, once aware of the transgression, assembles to confront the invaders. But there is no fighting. In each troop, which may number 15 to 20, males, females and young of all ages howl defiance at their neighbours. After a session of this both depart, having been assured by rules, at which humans can only dimly guess, that one troop was the winner, the other the loser. Doubtless, as in other animals that fight to preserve a boundary, it is the troop defending the territory that is spurred to the greater effort and is accepted as the winner.

amphibians, fishes and a few invertebrates such as dragonflies, butterflies and Fiddler crabs.

Territorial behaviour and the value that it has for an animal has been studied mainly in birds, whose songs and colourful displays make their movements easy to follow. The establishment and maintenance of the territory is usually the work of the male who advertises his ownership with songs and displays, the second function of which is to attract a mate, as will be described later. In this manner, the males of a species are spaced over the habitat and only those that gain territories obtain mates and breed. The spacing of the families reduces competition for food and other necessities such as nesting places, and lets them live in relative peace.

In his pioneer work, Eliot Howard concentrated on the territorial habits of warblers. These are small, drab birds which would easily pass unnoticed if it were not for their pleasant songs which fill the countryside on summer days. Howard watched male Reed warblers on their return from winter quarters in Africa. At first they went about in flocks but soon individuals slipped away for a short while to reed beds where they sang. Gradually, they spent more time away from the flock, until it no longer existed and each male Reed warbler had its own patch of reed bed. This was its territory which it defended against other Reed warblers by advertising with song and by chasing intruders. By this means, the birds learned the extent of their own and neighbouring territories; the intensity of singing and chasing then waned. The scene was now set for the arrival of the female warblers, which were enticed into the territories and courted. Once settled, the females also helped with the defence of the boundaries. They were, however, preoccupied with raising the family, so defence rested mainly with the males. The territories were maintained until the young became independent and then they were abandoned.

The holding of a territory confers a number of benefits upon a pair of Reed warblers. Eliot Howard thought that the territory provided all the food needed to raise the brood, in the form of insects and their larvae. The territory was seen as an exclusive larder without which successful breeding would be jeopardized. It is now realized that Reed warblers obtain most of their food from neutral ground outside the territories but, as we shall see, provision of an exclusive food supply is an essential feature in the territories of some birds.

Familiarity with the topography of the territory is a less tangible benefit. During the course of its daily activity a

Reed warbler gains an intimate knowledge of its living space. It learns the best places for finding food and the best route through the reed bed. Knowing where to look for food is a great asset when feeding a brood of hungry chicks and familiarity with the surroundings allows a bird to slip into the safety of cover when pursued.

While the idea of a territory as a home providing security and many material benefits holds for Reed warblers and many familiar songbirds, it has to be modified in one way or another for other species. Among birds of prey the territory is, even more than in Reed warblers, a source of food. It is a hunting ground, without which a bird cannot survive. When young Tawny owls leave their parents' territory in the autumn they must find an unoccupied piece of woodland in which to settle if they are to survive the winter. A private hunting ground is essential and many young Tawny owls die as a consequence of being endlessly chivvied by resident owls as they vainly seek a vacant niche in already fully occupied woodlands.

The young owls that die in their first winter are a **doomed surplus** whose fate is to die young. All animals produce more young than are needed to balance the number dying. They also form a reserve which will help to restore numbers when there is a catastrophe. If, for instance, a hard winter kills off many birds, their numbers will soon recover because the offspring of the survivors in the following year have a better chance of establishing themselves because of reduced competition. This, then, is another vital function of territory. It serves to regulate the population numbers of a species, so preventing overcrowding.

A clear demonstration of the regulatory function of territories has been given by studies of Red grouse feeding on the heather moorlands of eastern Scotland. Grouse spend most of the year in territories. The family parties split up in August, at which time many are shot, but by October pairs have set up territories again. During the winter a number of grouse die, but this is not simply a matter of food shortage. In the autumn young males compete for territories with their parents. Adults that previously held territory die soon after being ousted, not because they are denied access to food but because of stress. It seems that the loss of the territory is a shock that makes them lose condition and they quickly succumb. Some grouse gather in flocks on marginal ground, where there is little heather, which is their chief food. Some of these birds die of starvation but the majority are killed by

Invisible brick walls

Long before Eliot Howard had made his definitive study of the territorial instinct in animals, hunters had been aware of its principle. It is fair to suggest that prehistoric man may have learned of it without being aware of its implications. Certainly, since books on the hunt have been printed, hunters have commented how an animal being pursued will go so far then double back in its tracks, usually after having travelled in a wide sweep. This wide sweep represented part of the boundary of the territory or home range. So, all the experienced hunter needed to do, having chased his quarry so far and lost it, was to go back and wait, in the certain knowledge that the hunted animal would return.

In more modern times, the ornithologist, studying the territory of a particular bird or pair of birds, moves quietly and gently about in the vicinity of a bird's nest. By thus keeping the bird on the move, without alarming it, he can note the points where the bird doubles back or makes a wide sweep.

Another way is merely to sit still and note the bird's movements, whether on the ground or in the air. Occasionally the observer is rewarded by seeing the bird turn aside or turn back, as if it had encountered a brick wall, and it will repeat it at this same spot at other times.

predators. Their plumage, which affords good camouflage in heather, shows up well on grassland or in the fields of stubble where the outcast flocks take up residence. Whereas only two per cent of the territory holders are killed over the winter, seven times as many outcasts fall to predators. The predators, therefore, have taken the doomed surplus.

The possession of a territory does not automatically confer the right to breed. There has to be sufficient food. The grouse that have secured only a small territory will find it impossible to attract a mate, while those with the largest territories may pair with two females. The link between food, territory size and breeding is seen clearly in the Pomarine skua (or jaeger, as it is known in North America). This is a seabird which breeds on the Arctic tundra. During the breeding season it feeds almost exclusively on lemmings, small rodents which are famous for their cycles of abundance followed by a catastrophic decline into scarcity. The cycle follows a four-year period and when lemmings are abundant Pomarine skuas breed in territories of about eighty-thousand square yards. But when lemmings are scarce, the area of the territories has to be increased fivefold and the skuas may still be unable to breed.

Coots live together on the sur- face of a lake, but well spaced out. Every now and then the peaceful scene is disturbed by one male chasing another that has ventured into its territory.

Function of territory

The function of these territories is, therefore, to space out the species and to regulate the size of the population. At the same time, the environment is divided up according to its resources. Where and when food is plentiful, it can support a large population of pairs in small territories. A drop in food supply leads to deaths or emigration and the remaining animals redistribute themselves. In the woodlands, for instance, the densities of songbirds depend on the types of trees. In some Dutch woodlands there are areas of mixed broadleaved trees with thriving insect populations, interspersed with poorer pinewoods. In mixed woods such as these there were found to be four times as many Great tits, eight times as many Blue tits and twice as many Coal tits, as there were in pinewoods.

At the other end of the scale, many birds hold territories that play no part in the provision of food. This situation is typical among seabirds which feed at sea but breed in vast numbers on cliffs and small islands. The need for a feeding territory has been removed and the remaining function of the territory is to provide space for raising a family. The territory may be no more than a rough circle whose radius is the distance that a sitting bird can jab with its bill. Packing the nests together may be necessary because of limited space but some seabird colonies do not occupy the available ground. The inhabitants of these colonies derive a positive benefit from nesting in tight groups. The sight of a mass of their fellows indulging in courtship in some way stimulates others into breeding condition and the effect is to induce the whole colony to lay within a very short period. As a consequence, the chicks hatch out and fledge at the same times. By concentrating breeding into the shortest time, the chances of predation are reduced, since the closely-packed birds are able to present a united defence. This contrasts with the situation of birds owning large territories, whose single defence works on a different basis; by spacing their nests, they are less likely to attract the attention of a predator.

A third type of territory fails to provide room even for rearing a family. This territory is used solely for courtship. Among many of the grouse family, such as the capercaillie and Prairie chicken, as well as Birds of paradise, some waders (shore-birds) and other species, the males gather at a traditional display ground. Each male owns a small area, called a **court** or **lek,** where he displays to the other males. Typically, these males wear ornate plumage and advertise

Bowerbird artistry

A bower originally meant a dwellinghouse. Later the word was used for a shady place overhung with flowering vegetation or, more especially, for a fancy rustic cottage. It was the second of these meanings that inspired the common name of certain Australasian birds.

Not all bowerbirds build a bower, but all use a court or lek. In some species the male merely clears a small area of forest floor the size of a table-top, on which to display to his hen. Others go further and decorate their court with leaves or with coloured insect carcases, shells or flowers.

The male of the best known of the bowerbirds lays down a platform of sticks and in this plants two rows of vertical sticks. In the avenue so formed he runs up and down displaying to attract the hen. Other species may decorate an avenue with shells, bones and flowers or its walls may be painted with fruit juices.

The peak is reached in the bowerbirds of New Guinea that build a pile of sticks, roof this over with sticks, clear a lawn in front of it and surround this with a hedge of sticks, then decorate the garden so formed with flower heads.

Not surprisingly, the first Europeans to see these more elaborate bowers assumed they were the work of human hands, perhaps the result of children playing.

Keeping neighbours at wing's length

When the Common or Great cormorant returns from a fishing foray, it spreads its wings to dry them. That, at least, is the usual statement. Yet the cormorants will stand on

the rocks with wings held out when it is misty or raining. A cormorant may be seen to fly a mile to its usual rocks and, on landing, hold out its wings. A flight of a mile should have dried the wings. A cormorant may return to its rocky home and perch with closed wings, not spreading them for up to a half-hour later. None of this makes sense if the purpose of spreading the wings is to hold them out to dry.

An alternative suggestion is that this is a posture associated with territoriality. It could be a means of ensuring that in the event of a disturbance, causing them all to take off hurriedly and simultaneously, each bird has sufficient wing space to obviate the whole colony being thrown into confusion. Watching the birds more closely seems to confirm this. If one cormorant comes too close to another the latter will make an aggressive lunge at the newcomer, at the same time holding out its wings.

When we measure the distance between resting cormorants we find the spaces between them are never less than double the width of a spread wing.

their presence further with loud calls. By contrast, the females are drab. They come to the display ground, mate with one of the males and leave to rear a brood in seclusion. The role of **courtship territories** is discussed fully in the next chapter (p 132).

How a bird establishes its territory is not easy to see. Eliot Howard found that Tree pipits and whitethroats established their territories within the space of a very few hours and that the boundaries were later only slightly modified by border disputes with neighbours. On the other hand, Great tits and Blue tits seem to build up their territories by gradually exploring and extending their possession of an area. Once the territory has been established it has to be maintained against all comers. As soon as a male bird dies it is replaced either by its neighbours spreading their boundaries or by a new bird moving in. The defence of territories has been extensively studied. It is here that we see the expression of **conflict behaviour.** When confronted with a rival, an animal is torn by the conflicting urges to attack or to flee. When in its own territory it will be full of confidence and will attack, but on foreign soil it is likely to flee even if physically the superior. The confidence of the proprietor seems to be based on the knowledge that it is on its own ground; an intruder's willingness to flee prevents confrontations leading to actual combat.

In war, people fight more stubbornly in defence of their homeland than they do as an invading army. During air-raids on towns many people have experienced a greater confidence and ease of mind in their own homes than in the houses of neighbours.

At the actual boundary of a territory, where most confrontation takes place, overt fighting is reduced by ritualization of aggressive behaviour. The conflicting urges of attack and flight are expressed by displays, signals that communicate the state of motivation of an animal. An **aggressive display** may force a rival to retreat before the confrontation, and a display of **submission** by a defeated rival allows it to retreat without further punishment. Communication by these signals is discussed further in Chapter 11.

A special ceremony, the **triumph ceremony,** is found in all geese, which mate for life and have an intimate family grouping. Any intruder into a goose's territory is repelled by the gander, who then joins his mate uttering a special 'triumph note', with his neck stretched out and head low to the ground. The goose as well as the goslings all join in.

Male Kori bustard, of the African plains, in courtship display: transformed by fluffing the neck plumes and under-tail feathers.

Winter territories

It is little more than a half century since the concept of birds holding territories became firmly established. In that relatively brief period much research has shown how widespread in the animal kingdom is the territorial sense. Even so, this intensity of research, as is usual, has exposed many new problems.

The general idea is that a bird establishes a territory at the opening of the breeding season, which guarantees adequate living space for feeding and nesting. It has been found, however, that some birds also hold territories outside the breeding season. Among these are the European robin, blackbird and Great grey shrike and, in North America, the Plain tit, Loggerhead shrike and Red-headed woodpecker.

Why these should defend winter territories is hard to say. Perhaps it is a matter of getting in first, by staking a claim before the breeding season arrives. The Red-headed woodpecker may do so to ensure an adequate area for storing its acorns. The most puzzling of all is the European robin: not only does the male stake out a winter territory, but the female does too and she defends it. Yet when the breeding season draws near both male and female robins drop their exclusive territoriality and share a common territory. Robins' winter territories, also, seem to be unrelated to food supply since in times of shortage, as in hard weather, they do not hesitate to go outside their territories to feed.

Mammals' home range

The study of mammalian territorial behaviour is comparatively recent and it has been found that, among many mammals, the territory is not an exclusively owned, definite plot of ground. Most mammals spend the greater part of their lives in a particular area and this has often been referred to as their territory, although the area does not fit our definition of a territory. It is not necessarily defended and, rather than being an enclosed area, it may consist of a number of points of interest such as lairs, feeding places and pathways. It has, therefore, become customary to talk about a mammal's **home range.** Many young mammals lead a nomadic life after they have left their parents but, as they mature, they settle in one place and carve out a home range, which is defined as the area over which the animal normally travels in pursuit of its routine activities. Within this area it will have places where it will habitually sleep, a lair or a nest, favoured feeding grounds, waterholes, and so on, which are connected by paths.

As with a bird's territory, a home range is important in being familiar to its owner and conferring security. A mammal shows signs of unease if it leaves its normal range, but when at home it can travel round the pathways

Courage at home

There is more quarrelling between neighbours over boundaries than from any other single cause and, in Britain at least, more litigation. There is a parallel in the animal kingdom, for in spite of the old idea of males fighting over females the greatest source of aggression is in the maintenance of territorial boundaries.

Perhaps the other most interesting factor is that the possession of a territory automatically inspires greater courage. A follow on from this is that the nearer the owner of a territory is to the centre, where the nest is usually sited, the greater grows his courage. A territory owner may be smaller and physically inferior to an intruder yet invariably, or virtually invariably, he is the victor in any dispute.

Relations with man

Travellers in the South American forests have sometimes reported that they have been followed by jaguars for many miles. The jaguars make no attempt to catch up and in the end abandon the pursuit. The most likely explanation is that a jaguar views a man as a rival and is escorting him from his territory in the way that carnivores make sure that wandering individuals of their own species do not settle on their ground. This is also an explanation for instances of men being killed by large animals. They are seen as dangerous rivals, and unfortunately, cannot show their peaceful intentions with the necessary gestures of appeasement. Support for the idea is lent by the snarling, drawn back lips and flattened ears traditionally depicted for attacking lions and tigers. This is a sign of aggression towards another big cat; when attacking prey there is no expression.

blindfold. The traditional paths of badgers become beaten tracks that are so deeply rooted in the lives of the badgers that they continue to use them despite the upheaval caused by six-lane motorways being built across them. The result is that many badgers are run over, unless conservationists can deflect their movements through specially-placed culverts under the motorway.

From observations on captive animals, it appears that the first step in setting up a home range is one of exploration. The animal slowly makes its way round the unfamiliar area, making use of all its senses to find out as much about it as possible. Gradually, it establishes particular pathways and learns these, travelling more confidently at each circuit until it is quite 'at home'. So well learned are these pathways that the animal's feet land in exactly the same places every time it uses the path. This fact has long been appreciated by rabbit trappers, who position snares on tracks so that they can catch the rabbit in mid-bound. Small rodents, using pathways through the grass, put their feet in the same places and through this are aware of the precise location of every feature en route, including boltholes.

Within the home range, there are one or more **core areas,** where the animal concentrates its activities. These are usually near the centre of the range and include favoured feeding and sleeping places. Because of this, the boundaries of home ranges are difficult to define. The owners are rarely seen near the peripheries, and when neighbours meet they do not necessarily fight as do neighbouring birds. Neighbours can mix quite freely and it is not uncommon for home ranges to overlap. Among domestic cats, the home range consists of core areas connected by paths. Several cats occupy the same areas but they learn each other's habits and they can avoid meeting each other. Traffic at the crossing of pathways is regulated, the cats keeping a careful watch as they approach. If another cat is there, it is allowed to disappear around a corner before the second approaches. If two approach together, they sit and stare at each other until one breaks the deadlock and hurries across.

Complete tolerance of other individuals is seen in gorillas, which are docile animals quite undeserving of a reputation for ferocity. Gorillas live in groups and two groups may have almost identical ranges. On the other hand, the core area of a mammal's home range can be a territory in the strict sense. Lions defend a central territory, although they wander over a wider range. For other species, the whole home range is a defended territory. The vicuñas of South America live in groups of one male with several females. The male prevents other males from entering the territory.

Marking the boundaries

Stoats, weasels and other members of the Carnivore family Mustelidae, which includes martens, mink and otters, hold defended territories which appear to form the whole of their home range. These territories are hunting grounds and are related in size to the abundance of food. The territories of stoats are eight times the size of those of weasels although stoats need only twice as much food. Weasels are small enough to hunt mice and voles down their burrrows, whereas stoats must hunt above ground and so need more space. The boundaries are patrolled regularly and, as a form of 'beating the bounds', they leave **scent marks** of urine and droppings, with secretions of the anal glands added. This form of communication, which is the equivalent of the domestic dog marking lamp posts, is discussed in Chapter 11 (p254). Within the territory, the stoat or weasel has a movable core area. It hunts in one place for a few days, then moves on to the next. However, it does not ignore the rest of the territory and is continually going out on patrol. This is necessary because if a territory is left unattended (perhaps through the death of the owner), a new animal very soon moves in.

The new animal would be a **transient,** a young animal not in breeding condition. Indeed, it cannot come into breeding condition until it occupies a territory and so it tends to wander without settling long in any one place. Like the young Tawny owl, it is a nomad and is chased from territory to territory by the respective owners. This means that it cannot gain the intimate local knowledge needed for successful hunting. If found by the occupant, the transient is escorted from the territory in a mad chase in which both animals abandon their usual care for keeping out of sight.

The description refers to the behaviour of male mustelids. Females hold smaller territories within those of the males. The boundaries of **female territories** do not adjoin and the female defends her territory mainly against the male in whose territory she lives. She is subordinate to him except when she has a family. Also, she stops defending her territory when she is in breeding condition. Among the coatis of tropical America, groups of females and juveniles defend territories and adult males are rigorously excluded except during the breeding season.

Committee rules
Verheyen, the Belgian zoologist, spent most of his professional life, in the national parks of the Belgian Congo, now Zaire. Among other things he made a very complete study of the hippopotamus. In the course of this he found that the females and juveniles occupied a central area of the river, normally with a sand bank where they could haul out and sleep. Each male had his territory on the perimeter of this area: the older and therefore dominant males had territories contiguous with it, the younger and subordinate males farther from it.

A female in oestrus could enter the territory of a dominant (i.e. breeding) male, where quite naturally she would be received amicably. At other times, a female leaving the home area, especially if she had a calf at heel, was liable to be attacked on meeting a male. A male could enter the central area where the females and young were normally quartered, but only if he obeyed certain rules. Thus, if one of the females, who usually would be lying down when he entered, stood up, he had to crouch to the ground, or in the water.

In this way, the females could completely govern the movements, and the extent of those movements, of a visiting male. Should he break the rules he would be set upon by several females at once.

This is like committee procedure, with the unwritten law that if the chairman rises to his feet a speaker must sit. It is an obvious way of preserving peace and order.

Female coati, with three young, defending her territory against an intruding male. Coatis, near relatives of the raccoon, live in the forests of the warmer parts of America.

As with some birds, there are mammals which in the mating season assemble in small **breeding territories** with the exclusive purpose of mating. The antelope known as the Uganda kob is a well studied example, which is discussed at the end of this chapter, and there is a parallel between the seals and seabirds which feed at sea and come ashore to breed. For both seals and seabirds there is no need for the territory to provide food. The Fur seals of Arctic and Antarctic regions gather to breed on traditional beaches. The bulls arrive first and stake out territories by fighting and threatening. The largest bulls get the best sites, which are on the open sand where the cows congregate to bear their single pups. Smaller bulls find territories inland of the beach or along the tide's edge. At high tide, a row of young bulls can be seen up to their necks in the sea. Although the mature bulls attempt to round up the cows, the latter are free to come and go, so a bull's success in mating depends on his gaining a patch of ground favoured by the cows. The territories are quite small and have well-defined boundaries. It is possible for a man to walk along a boundary in safety because the bulls range up on each side, none daring to attack the man since by coming up to the boundary it would provoke an onslaught from its neighbour.

113

In general terms, therefore, it is preferable to speak of a home range where mammals are concerned, rather than a territory. Even so, with those that have a fixed home, and this naturally must lie within the home range, there may be a small area, the core area, around it which is defended against intruders of the same species. This could be, and is by some authors, treated as a territory in the strict sense. Many small rodents probably have territories but because they are nocturnal it is difficult to make the necessary observations to set the matter beyond doubt. Another source of doubt is in those mammalian species that have a **social hierarchy** (see Chapter 10). Often there are fights between the dominant individual and a subordinate in the peck order, and this may have all the appearance of a border fight when in fact it is well inside the border.

Some mammals indicate the boundary of a territory or a home range by **vocalizations,** as birds do. Troops of monkeys have been seen to approach each other at what was presumably a common boundary and threaten with much chattering and grimacing. Howler monkeys, of South America, use their much-enlarged voice box in noisy howling contests under these same circumstances. The males play the major role in these vocal battles but the females also join in, even the juveniles, although they do no more than whine and this may be more from apprehension. Howler monkeys also have howling sessions at dawn and dusk and this is sometimes interpreted as being a means of advertising to other troops that they are in occupation.

Usually mammals mark boundaries with an odoriferous substance such as dung or urine, as in dogs and hippopotamuses; others have special **scent glands,** as in civets and Musk deer already mentioned. Deer and most antelopes have special scent glands on the face and legs or between the hoofs. Only within the last few years has the **chinning** of rabbits been understood. They rub their chins on sticks, grass, pebbles and the like to distribute a scent given out by glands under the skin.

Visual marking of boundaries by mammals is not common. It may be used by monkeys and their boundary vocalizations could then be interpreted as reinforcing this. The kongoni, an antelope of East Africa, is known to use a bare patch of ground, known as the **stamp,** from which the male leader of the herd warns possible intruders by ritualized head shaking and mock charges. Any intruder persisting in transgressing the boundary will be met with a head-on clash.

Closing in on the enemy
A competent naturalist once pointed out to us where a fox, the only wild fox he knew to have died of old age, used to live. The place was right on the edge of a kennels, in Devon, England. The naturalist made a charming little drawing of a fox with a woollen scarf round its neck standing in a queue at the local post office to draw its old-age pension.

The story illustrates a fundamental principle in zoology, that predators do not hunt in the vicinity of their core area. In this case the pack of hounds and the huntsmen would set out but the hunt would not begin until it was some distance from the kennels.

A fox has been seen to set out on its daily hunt and pass a pheasant at close range or play with a leveret (young hare) before going off to hunt leverets.

Small birds that are prey to raptors (birds of prey) have often been reported as having built their nest in a tree underneath the nest of a hawk or eagle. They have raised their families in what looks like a most dangerous position, yet have never been molested, simply because they are sited near the predator's home base.

114

Stories of toads found in cavities in rock or wood are legion, implying that they live for hundreds of years. The record for longevity for the European toad concerns one that lived for 36 years under a front doorstep and would no doubt have surpassed this had it not been killed by a tame raven.
The more remarkable feature is that it should have remained so faithful to this one resting place. But toads are like that. One spent many years under a flower pot in England. At almost any time of day, and every day except in spring, one could lift the flower pot and see it there in the same position.
There is something stolid and reliable about the European toad, epitomizing almost the hallmark of good citizenship. It does not wander far from home, being able, it seems, to subsist on insects caught within a fairly limited radius of home, largely by walking ponderously around and waiting for its food to come to it. It is the more remarkable therefore that it should, at times, make long journeys. One of these times is the breeding season, when the toads proceed slowly, doggedly and surely to their breeding ponds, a mile or more away, returning faithfully each to its separate home when this is all over. Several species of toads have been tested experimentally and found capable of returning home when taken a mile away and then liberated. It can only be suspected that, like migrating birds, they use celestial navigation, guided by the sun by day and the stars by night.

Familiarity with garden songbirds leads to the mistaken impression that all territories are owned by a single animal. Among mammals, this is often not the case. Prides of lions, parties of coatis, troops of baboons and packs of wolves defend **communal territories.** Even rats and mice live in groups or 'clans', and recent studies show that they should be considered as members of a fairly rigid society. Members of a clan are identified by odour; a strange mouse is attacked until it has acquired the communal odour by mingling with the clan.

A clan of Wood mice, like House mice and the American Deer mice, lives in the territory of the dominant male. The territory measures some four to six acres and its boundaries are defended by the male. Like any military leader, the male patrols his territory confidently and his subordinates have to keep clear of him. Any that interfere are driven away or killed. This despotic but stable society is important in regulating numbers of Wood mice. Young males are driven out and will die unless they can set up a small range where they can feed without meeting a dominant congener. Within the territory, each individual has a home range of less than half an acre which overlaps those of other individuals and contains several nests.

A number of birds hold communal territories, but without the individual home ranges seen in mice. Mexican jays live in small flocks of about a dozen birds which consist of adults and yearlings that have stayed with their parents. The flock defends a communal territory against other jays and they also mob (harass) predators. Each adult female builds her own nest and incubates her own eggs. The males and juveniles combine to feed the sitting females and, later, the chicks.

Homes for cold-blooded animals

The discussion of territorial behaviour has, so far, been about mammals and birds but a **territorial instinct** is found in other groups of animals. Many species have a sense of 'home' in that they tend to stay in a restricted area and to return to a particular place of refuge. This sense is particularly well developed in the limpets: as the tide falls, each animal makes its way to a particular place on a rock. The limpet's shell has grown so that it fits perfectly the irregularities in the rock surface of its 'home', thus forming a watertight seal when the limpet clamps down. Then, when the tide rises again, the limpet crawls over the rock, feeding on the adhering layer of seaweed. If one limpet meets another limpet, a stranger, on its feeding ground, it feels the stranger with its sensory tentacles then pushes against it until it retreats.

How a limpet finds its way back to its 'home' is not known. It has been suggested that a limpet uses the Earth's magnetic field to home. The Common toad, of Europe, also seems to have a preferred resting place, a damp secluded place where it is safe from drought and enemies. Often it lies up under a log or boulder where it retreats after each night's foraging. One toad lived for thirty-six years under the doorstep of a house. A toad probably uses **celestial navigation,** using the sun by day and the stars by night.

Territory and range in the broad sense is developed in some **insects.** The male bumblebee patrols a circuit of a few hundred yards. He stops at intervals to mark leaves and twigs with a scent that attracts queen bumblebees. Several males have overlapping circuits and their continual patrolling soon leads to a meeting between one of the males and a queen attracted to the area of the scent. Male dragonflies patrol stretches of river or road with the same end in view but their beats are proper territories, interlopers being driven away. Their method of maintaining the territory is to attempt to mate with every dragonfly that approaches. Females accept the courtship but males flee. Certain butterflies hold mating territories, and in one species, the butterflies gather on a hill. The dominant male seizes a territory on top of the hill, with subordinate males occupying literally lower positions.

A number of **fishes** are territory holders. Three-spined sticklebacks, for instance, are sociable and live in small schools until the spring, when the male changes from having a green to black back to a bright green back with red breast and throat. The red in particular is a signal to other sticklebacks that he owns a territory. His territory is an area of river bed from which other males are driven. He rushes at intruders with spines erect and mouth agape. The intruder flees, unless it is at the boundary of its own territory, in which case it stands on its head and jerks up and down. The defender follows suit and the two display to each other and perhaps fight. More likely, they eventually retreat, having clearly settled where the boundary between their territories lies. Female sticklebacks are silvery, lacking the red coloration of the male but become swollen with eggs. Only a gravid female is welcomed by the male. He vigorously courts

Limpets have a strong attachment to 'home', a circular depression in a rock cut by the edge of the shell.

Two male Three-spined sticklebacks meeting at the boundary between their territories display aggressively with spines erect and fins spread.

Belligerent butterflies

It seems almost preposterous to speak of a butterfly as aggressive, yet there is one that is always described in this way. It is the Small copper butterfly.

The belligerence of the Small copper springs from the male establishing a territory and trying to chase out all other butterflies, flying out and attacking individuals of its own species as well as those of larger species.

The Comma butterfly also shows territoriality. It is a European species, 2 inches across the spread wings, red with dark markings on the upper surface of the wing. Apparently it only defends a roosting territory. When the sun is going down, and feeding is over for the day, a Comma butterfly, named for the white comma on the underside of each hindwing, takes up position on a tree. Where two Comma butterflies have adjacent territories each will fly up and challenge his neighbour should it intrude. Even a dying leaf, the colour of the comma's upper surface, fluttering down from the tree overhead, will be attacked.

each gravid female that enters his territory and it is he who tends the eggs and later the young sticklebacks. He relinquishes his territory only when the young sticklebacks disperse.

Among sea fishes, cod form territories on their breeding grounds. They gather well above the sea floor and males manage to maintain territories without obvious boundaries. Unlike the stickleback, a female cod lays large numbers of eggs which float freely in the sea, so the male cod's need for a territory disappears after mating. Territories similar to those of the stickleback are maintained by cichlids.

There are six hundred species of cichlid fishes living in the warmer parts of the world. They are popular as aquarist's tropicals. As the breeding season advances, the male's coloration intensifies and he becomes less sociable. When any fish, particularly one of his own species, swims into his territory, he immediately goes into display, erecting his fins and raising his gill covers to show off his bright colours. The response of a female cichlid is to react coyly, thus establishing that she is there to be courted rather than to be fought. Males out of breeding condition flee, but intruding males in courtship dress are belligerent. However, the territory-holder appears to have the psychological advantage of ownership, and usually wins an encounter. The advantage lasts only while he is in his own territory. The boundary of the territory is invisible to human eyes but, to the cichlids, it is as defined as a brick wall. The owner will dash at an intruder, then stop dead as if he had collided with a pane of glass. He has reached the boundary and will go no farther. The territory is maintained throughout the breeding period and acts as a refuge for the family of baby cichlids which stay with their parents. Other cichlids recognize the boundaries and do not normally cross the threshold.

Among the **amphibians,** territorial behaviour has not been widely studied. Many frogs, toads, newts and salamanders gather in the breeding pools and sometimes form dense masses with several males jostling around one female; they do not hold territories. Only a few species, such as the North American bullfrog and Green frog, are known to hold territories. Males of these species keep well spaced-out in the breeding ponds, presumably advertising their positions by calls which attract females and repel males.

Many **reptiles** are gregarious during the breeding season. Rattlesnakes gather to hibernate in dens where scores may be found intertwined. They mate on emergence in spring, before dispersing. Lizards, on the other hand, are often strongly territorial. The males don a bright courtship coloration and defend well-defined territories. During morning and afternoon, when the sun is not too hot, the Rainbow lizard of Africa, one of the agama family of lizards, can be seen defending its territory, sometimes on the wall of a house. When it meets a rival, the two lizards display the coloured gular fold under the chin and bob up and down in comical jerky press-ups. Occasionally, a fight breaks out and the pair stand head to tail alongside each other and lash out with their tails. In particularly fierce encounters, tails and jaws may be broken. Agama territories are held throughout the breeding season, and each lizard gathers a small harem of females. However, there is a truce every night. As the sun sets, the males leave their territories and gather in a communal roost up a tree or under eaves of a roof. Tolerance is assured by each lizard going dull brown. The breeding colours are resumed the next day.

Dragonfly patrol
The Small copper butterfly and the Comma (p 117) are not alone among insects in establishing a territory. Dragonflies regularly do so, each male patrolling a well defined air space over a pond or stream and flying at any intruding male (see illustration). Here, however, we are dealing with a carnivorous insect with jaws capable of inflicting some damage. After skirmishes of this kind dragonflies often show signs of their fighting in the shape of torn wings.

Fallow deer buck (top) guarding his harem, doing battle (right) with a potential usurper of his territory, and (below) chasing his defeated opponent from the territory.

Few fights over females

Had the idea of a territorial sense been substantiated earlier it would have robbed the novelists and the nature-writers of a fruitful source of plots. As it was, there was overwhelming acceptance that when two males, human or animal, fought it was over a female. In practice, from the few data available, it is more common for two females to fight over the possession of a male.

Even when Eliot Howard's work on territoriality in birds had become widely known the old idea remained more or less intact. Full realization came when K. H. Buechner's film of the Uganda kob became available to scientists. In this film we see the male Uganda kob patrolling a small area of ground, his breeding territory. Around this territory, but not contiguous, are several other territories, each with its patrolling buck. These are all separate breeding territories within the home range.

Into the first territory comes a doe. The male takes no obvious notice of her, nor she of him. The doe settles on the ground in a resting crouch. Very soon another doe enters the territory and settles down beside the first. The male continues his patrolling.

There is occasional fighting between the bucks of the Uganda kob, but only when one on the way to and from his breeding territory and the feeding ground inadvertently crosses another buck's boundary. From Buechner's film and his written account it is clear there is no fighting among males over possession of a harem and it is obvious that the does choose their mates. Once a doe has entered a territory there is a tendency for other does to join her, rather than enter a territory occupied by a buck only, as if the newcomers are relying on the first doe's choice.

Red deer and wapiti form harems at the rutting season and for each harem there is a dominant stag that charges endlessly around the harem, wearing a gully in the earth with his hoofs. The usual written accounts of the Red deer rut give the impression that the stag rounds up as many hinds as he can and fights any other stag coming near for possession of the hinds. Once the precise situation regarding the Uganda kob is appreciated, it becomes clear that the stags are not fighting over the females but over territory. This is reinforced by what is known from observation of the Red deer harems themselves. The hinds accept the domination of the stag temporarily but retain the normal matriarchal social structure within the harem. Should danger force them to

119

take flight, the stag makes off at full speed, leaving the hinds to look after themselves.

Whenever two male animals are fighting over territory there is almost bound to be a female nearby. She is seeking or has found a mate. But the appearances, which we now know to be deceptive, are of the males duelling for the fair lady's hand, when all we are seeing is a violent quarrel over a piece of real estate. Seldom, if ever, do males 'fight for possession of the female', as is so frequently asserted.

Chapter Six
Starting a Family

asexual reproduction: binary fission: budding: parthenogenesis —
sexual reproduction — sex attractants — oestrus: anoestrus —
polygamy: polygyny: monogamy — courtship: courtship display:
courtship feeding — hierarchies and choice of mate: leks: rutting
grounds — pair formation — reproductive strategy — mating —
ensuring fertilization — sexual roles — hermaphroditism — sex
reversal

Perhaps the most characteristic property of living things is the power of reproduction. The individual owes its life to the perishable nature of its body and sooner or later succumbs to the destructive influences of its environment. Before yielding to the inevitable it will, under normal conditions, have produced offspring which will carry on the struggle for another generation. As the epitaph on one headstone in a graveyard has it: 'We are born to live, to die. Why?' The eternal mystery in this brief statement has yet to be solved, but one inescapable element in it is that one part of each individual, as a rule, lives on, to repeat the cycle.

Fundamentally, reproduction is a factor of growth. It can be traced back to the division of a simple ancestral mass of protoplasm into two or more parts whenever its volume increased to the extent that the surface was no longer sufficient for the necessary intercourse between the organism and its environment. Following this division the original proportions are restored, by feeding and growth, and once again division takes place. So, while everything exists for its own benefit it also exists for future generations.

Only in the simplest organisms is this basic expression of life obvious. For the rest it has become overlaid with a multitude of activities and emotions that obscure the original

She does the courting
Roe deer, of northern Eurasia, are small, up to two and a half feet at the shoulder. They live in woods adjoining grassland, mainly hiding by day, coming out at night to feed. They do not form herds but live solitary, in pairs or in family parties while they have young. Roe do not readily show themselves but their presence is sometimes given away by the roe rings or racing rings, well worn tracks in the ground which may be irregularly serpentine, may form figures of eight or, typically, circles. Usually the circles have a tree at the centre, a bush or a tussock of grass, and may have a radius of a few feet to twenty feet.
Those who have watched the rings being used report that the doe catches the attention of the buck who follows her nose to tail as she dashes off, he hard behind her emitting a pant-

ing sound. The chase may be fast, or the pair may walk. For the most part the buck seems to be heading the doe off, so driving her in a circle, or he may shoulder her from one side. In all cases there is the appearance of the buck dragooning the doe, as if she were an unwilling partner in the courtship game. For a long time it has been debated what the purpose of these racing rings might be. That they are part of a courtship ending in mating seemed certain and several improbable theories have been put forward to account for the roe's behaviour. The current view is that it is the doe who entices the buck, the running around stimulating him to mate.

source, activities such as courtship and mating, emotions of love, affection and hate, much of which must be left to the moralist to explore and the poet to describe. The biologist needs all his time to unravel the pure mechanism.

Reproduction, in its most familiar form, can be conveniently divided into two stages. The first involves the finding of a mate followed by copulation or coition. The second starts with conception, and ends when the young become fully independent and leave home. In this chapter, the first stage only is considered. It may be a brief, no more than an ephemeral joining of the sexes, or it may be prolonged, involving a period of courtship extending over several years. Much depends on which level of the animal kingdom is under review.

It is even possible for the first stage of reproduction to be omitted altogether, as in the ancestral mass of protoplasm already referred to, or even in higher organisms which will be mentioned later. This simpler process is known as asexual reproduction, in contrast to sexual reproduction, which is what we normally are referring to when we speak of reproduction. In this, two sex cells, or **gametes,** the male sperm and the female ovum, fuse to form a **zygote,** the fertilized egg. The zygote then develops into an embryo and ultimately the adult individual. Courtship, pair formation and copulation all have as their ultimate purpose the simple phenomenon of bringing together a sperm and ovum. However, as we shall see, there are instances throughout most of the animal kingdom of asexual reproduction in which there is no union of sexes and no fertilization. At its simplest the animal reproduces by binary fission, a mere splitting in two.

Asexual reproduction

Binary fission Among the single-celled protozoans, binary fission is the main method of reproduction. It is exemplified in amoeba, the textbook protozoan, so familiar that its scientific name has now become its common name. All the functions of life of amoeba are carried on within a single unit of living matter only a fraction of a millimetre across. Although it has been watched for more than a century, amoeba has never been seen to reproduce by any means other than binary fission. At a given point in time, an amoeba begins to change. The nucleus starts to pull apart into two halves, so that it is shaped like an hour glass. At the same time, a constriction appears across the body, as if a fine thread were being tightened around it. This gradually deepens and eventually the amoeba splits in two, each part containing half the original nucleus. Each part grows and eventually both will also split into two. *Chlamydomonas*, a protozoan that swims by means of two protoplasmic hairs known as flagella, goes further. It may divide once, twice or three times to produce two, four or eight 'daughter' cells. Sometimes the new cells remain within the protective membrane of what remains of the 'mother', where they can survive periods of adversity.

Budding Another form of asexual reproduction is seen in the budding which occurs in hydra, the familiar classroom relative of Sea anemones. Hydra reproduces sexually as well, but when conditions of growth are favourable, presumably when there is a superabundance of food, it develops buds on the surface of its tubular body. Over the space of a couple of days, each of these grows out from what is no more than a bump on the outer surface into a miniature hydra, with a hole forming the mouth at the tip and a ring of tentacles around it. Finally, it is nipped off at the base and becomes independent of the parent. This baby hydra floats away, then settles down to grow into an adult. Several buds may form at one time on different parts of the body.

Sea anemones have an even greater propensity for asexual reproduction. An anemone may split into two longitudinally or it may form a new ring of tentacles halfway down its body, after which the top half breaks off to give two anemones in the place of one. Some species of Sea anemone form buds, rather as in hydra, but around the base of the body. Others crawl over rocks, pieces being torn off the base of the body as it moves along, each piece growing into a minute but perfect anemone.

Parthenogenesis Asexual reproduction is also found among the higher animals. Here, the egg develops into an embryo without fertilization and thereafter development proceeds as normal. This process of virgin birth is called parthenogenesis (Greek *parthenos* = virgin). It is fairly common among insects, notably among the aphids or Plant lice. Most of the year aphid populations consist solely of wingless females. These are the familiar pests of crops such as beans, and of rose bushes. They spend the whole of the summer season bearing a host of young parthenogenetically; but late in the year winged females are born that lay eggs which hatch into males and females. These mate, and the females lay eggs that are fertilized and survive the winter to hatch into wingless

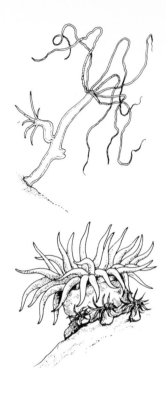

Reproduction by budding is very common in animals of the phylum Cnidaria: these include Sea anemones, jellyfishes and corals, as well as the freshwater hydra. Hydra gives off buds from the sides of its body; some Sea anemones give off buds from the base of the body.

parthenogenetic females the next spring and so the cycle starts again.

The main advantage of parthenogenesis is that it gives a very rapid rate of breeding. Swarms of aphids can appear almost overnight as there is no delay between generations for mating. The change from live births to egg-laying is controlled by temperature and in warm greenhouses aphids produce generation after generation of females without cessation and with never a male. The Crustacea contain quite a number of species in which the females produce eggs that do not need to be fertilized. Among them are the Water fleas which under favourable conditions are all females producing eggs that give rise to other females without being fertilized, this process being repeated for hundreds of generations. Only when adverse conditions threaten, such as the drying up of the water in which they are living, do males appear among the offspring. Then some of the females produce eggs that need to be fertilized. Such eggs can withstand adverse conditions and so carry the species on through a period of difficulty.

A similar thing holds for a few reptiles such as the Chequered whiptail lizard of Texas and *Lacerta saxicola*, a lizard of the Caucasian mountains. Large scale collecting of the latter revealed that whole populations exist without a single male. There are bisexual populations of these species, however, where mating and fertilization take place.

A favourite of aquarists is the Amazon molly, a small black fish from the rivers of Mexico and Texas. It is a natural hybrid from matings of the Sail-bearer molly and the Sail-fin molly. It reproduces parthenogenetically and only one offspring in ten thousand is a male. The odd thing is that the Amazon molly has to mate to produce offspring. The female mates with a male of any other molly species but the offspring are identical with her. The act of copulation seems to stimulate the development of eggs even though the male's sperms do not fertilize the Amazon molly's eggs.

Sexual reproduction

Although asexual reproduction occurs intermittently well up the animal scale, the first signs of sexual reproduction are seen as low down the scale as the single-celled protozoans. *Chlamydomonas*, already mentioned, not only reproduces asexually by fission, but may also produce gametes which fuse to form a zygote, a process called **conjugation** because the gametes are similar in size, with no differentiation into male and female. Individual *Chlamydomonas* merely lose their cell walls to become gametes. They then form clumps which later disperse in pairs of gametes, each pair fusing at their front ends to form zygotes. Each zygote secretes a thick protecting wall, which, about a week later, ruptures to release four 'daughters' that have been produced inside the protective wall, by simple division, and these later multiply by binary fission. In another protozoan, *Vorticella* or the Bell animalcule, there are the first signs of a division into sexes. Conjugation follows a pattern similar to that in *Chlamydomonas* but one of the gametes is smaller and more active than the other. The larger plays the part of the ovum of higher animals while the smaller represents the mobile sperm. Generally, this distinction of the male being the active partner is retained throughout the animal kingdom, the disparity between the sexes being known as **sexual dimorphism**.

In view of the fact that even some advanced animals reproduce asexually, we may wonder why the more complicated procedure of sexual reproduction ever evolved. The answer is that sexual reproduction introduces variety into a population. In asexual reproduction, the offspring will be genetically identical to the mother but in sexual reproduction the fusing of gametes introduces characters from each parent into the new individual. Furthermore, it ensures that all the gametes from one parent will differ slightly from each other so there is the potentiality for an infinite variety in the offspring. This variety is fundamental to the process of evolution by natural selection. The pressures of natural selection are continually weeding from the infinite variety those individuals that are unsuited for prevailing conditions. Those that survive reproduce and pass on their suitable characters. If, however, conditions change, individuals with a different set of characters may survive. Thus, a population of parthenogenetic lizards may continue successfully until there comes a crisis, such as the appearance of a new predator. If they cannot cope they will be wiped out, whereas a sexually produced population may number among its varied individuals a few with better camouflage or a faster turn of speed. These few stand a better chance of survival and of passing on their particular advantage to their young who soon make up the entire population.

Recognition of a mate A basic necessity for sexual reproduction can be summed up in the word recognition: an animal must be able to recognize that a potential mate

Giant sperm
The usual shape, of a sperm (male gamete) resembles that of a tadpole with a rounded or oval head and a long tail. Those of some crustaceans are merely rounded or may be star-shaped.

There is a subclass of the Crustacea known as the ostracods in which the body is enclosed in a bi-valve shell. Most ostracods are no longer than 1-2 mm and the largest is no bigger than a cherry stone, but they have the most fantas-tic sperms in the whole of the animal kingdom. These sperms have long heads and may be up to ten times the length of the male that produced them. In fact, these diminutive crus-taceans, most of them barely the size of a pin's head, produce sperms that are longer than those of an elephant or even the largest whale.

It has been suggested that these sperms cannot be functional and when we link this with the fact that the majority of ostracods reproduce by par-thenogenesis, that is, the females lay eggs that do not need to be fertilized, we begin to wonder what it is all about.

From time to time the student of zoology comes across remarkable in-stances of this kind in which there seems to be neither rhyme nor reason. In this instance it may well be that some day somebody will discover the reason and unfold a quite spectacular story but for the moment one can only regard it as evolution gone crazy.

Benign deception

Most fishes lay their eggs as if they were sowing seed, leaving them to their fate and taking no further trouble over them. A few make a nest and care for them by protecting them from predators as well as fanning with their fins to drive oxygenated water over them. In between are the mouth-brooders in which either the male or the female, rarely both, takes the eggs in its mouth, keeping them there until they hatch. This not only protects them from predators but also ensures a constant flow of water over them, the water the parent takes in for respiration.

Whichever parent broods the eggs in the mouth must snap them up as soon as they are laid. This gives little time for the eggs to be fertilized, and has led to a very neat device for ensuring fertilization.

There are a number of species of cichlid fishes in African lakes which are mouth-brooders. At the start of the breeding season the males put on bright colours, a not uncommon event in fishes, but they also develop spots on the anal fin. These were called ocelli or eye-spots, at first, such as many small animals carry to scare away predators. The typical eye-spot has, like the eye, a dark centre. Those on the male cichlid have none and in size, shape and colour they resemble the cichlid eggs. Then it was found that the female cichlid snatching up her eggs also tried to snap up her mate's 'eye-spots'. This brought her mouth near his genital opening so that she also snapped up the sperms her mate was emitting. So her eggs were fertilized in her mouth.

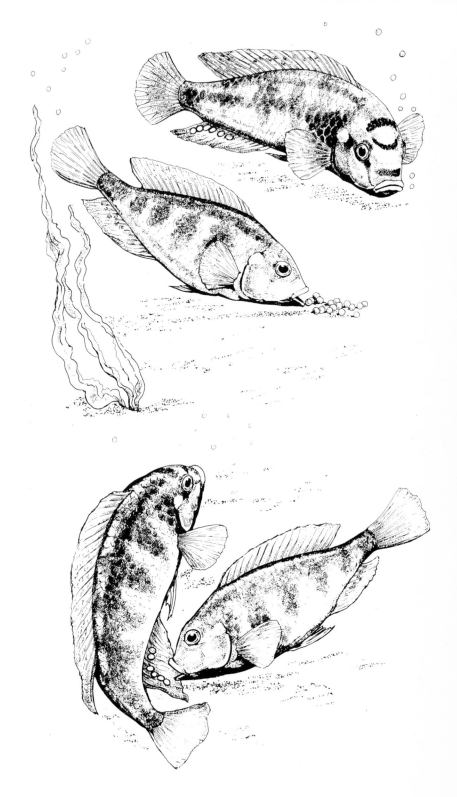

127

belongs to its own species and that it is of the opposite sex. The importance of recognition is seen particularly among spiders where the female is likely, in any event, to treat the male as prey. Recognition is based on the use of **signals**, often ritualized, that indicate unequivocally the species and the sex and, very often, a willingness to mate. The signals are conveyed through the medium of virtually every kind of sense organ. Exchange of these signals constitutes what is called **courtship**, the preliminaries to mating, which can vary from the transmission and reception of one simple signal to an elaborate, sometimes extravagant, sequence of posturing and calling that may take days or even weeks to reach its culmination.

Among the lower animals, reproduction may not involve actual copulation. Aquatic animals release their sperms into the water, there to make their way to the ova, which may be retained in the females' bodies or may be liberated into the water. The sperms are attracted by chemicals but, to ensure fertilization, the liberation of sperms must coincide with the ripening of the ova. In the Palolo worm (p 35) synchronization is controlled by phases of the moon, but more usually it is effected by some form of signal. *Chlamydomonas*, the simple protozoan, gathers into mating clumps through the stimulus of a **pheromone** released into the water from the flagella. The pheromone from one **clone** (a group of individuals descended by asexual reproduction from a single ancestral 'mother') attracts and causes clumping only in members of another clone. Thus inbreeding is prevented and mixing of characters is ensured. However, in most cases it is not known how liberation of sperms is controlled nor how they are attracted to the ova.

Chemical signals are particularly suitable for bringing together the sexes. No elaborate transmitting or sensory organs are needed and the chemicals diffuse speedily and comprehensively through air or water. Among animals that have been studied, pheromones are employed in particular by insects and mammals. For an example of a mammalian pheromone or **sex attractant** we need go no further than that which attracts the neighbourhood's dogs.

The pheromones of insects have been widely studied because of their use in the control of insect pests. Attempts have been made to eradicate pests by putting out traps 'baited' with the female sex attractant. The male insects so caught are then killed or sterilized. An example is the Gypsy moth, a European species that was introduced into the

Sex appeal in crabs
A large part of our knowledge of animal behaviour has been contributed by scientists who spend a long time studying one particular species. One person may spend weeks, months or even years on one species until they almost get to know the various individuals in the study area as old friends.

One animal studied intensively is the Fiddler crab, so we know a great deal about its courtship and its general habits. We know how the males and females pair off and we also know that in any large colony of Fiddler crabs there is likely to be one female who, the moment she runs across the sand, catches the attention of all the males. Even those males that are at that moment courting another female are likely to break off and give attention to this one female. The same thing happens in cuttlefishes. There may be a large number of cuttle-fishes going about their various occasions, when one female swims into view and becomes the cynosure of all the males. Even those indulging in a mild flirtation elsewhere will be attracted to break off their philandering and give their attention to this one female.

Any scientist who takes time off to study a colony of human beings would be able to report the same findings.

eastern United States from Europe in the late nineteenth century. Its caterpillars destroy the leaves of trees and it soon became a pest. The female moth cannot fly and has to wait for the male to seek her out. To this end she liberates a pheromone that attracts male Gypsy moths only. The pheromone was identified by extracting minute quantities of it from several hundred thousand female moths. It is extremely powerful and as little as one ten-thousandth of a millionth of a gram is sufficient to attract a male. A synthetic pheromone, made from castor oil and called 'gyplure' has been used successfully as a means to combat Gypsy moth infestations.

Another form of 'trap', based on vibrations, was discovered accidentally by Hiram Maxim, the inventor of the Maxim machine gun. In 1878, he erected a line of electric street lamps to illuminate the grounds of a New York hotel, and noticed that mosquitoes gathered each evening around the transformer. Being an inquisitive man, he took a closer look at the mosquitoes and recognized them as males by their feathery antennae, those of females being club-like. They were attracted by the hum of the transformer, mistaking it for the hum produced by the beating of a female mosquito's wings. The males detect the hum with their antennae and are attracted over a distance of ten inches. Male mosquitoes are not attracted to each other because their wings beat too fast. They respond only to frequencies of 300-800 cycles per second, the female's wingbeats being 450-600 per second. Moreover, newly emerged females have a slow wingbeat and they begin to attract males only when they mature and their wingbeats speed up. Young males do not attempt to mate because the sensitive hairs on their antennae which 'trap' the sound waves do not unfurl until they mature.

Readiness to mate Once another animal has been identified as of the same species but opposite sex, mating can proceed; but first it is necessary that both partners should be in a suitable state. Mating is pointless if the gametes of either partner are not ripe. For the mosquito, as we have just seen, there is a simple, automatic signalling system based on physical changes to the wings and antennae. Among other animals there are complex systems in which signals from one individual affect the physiological state of another. We have seen how the displays of the male canary (p 25) stimulate the secretion of sex hormones in the female to bring her into breeding condition and in rabbits the act of mating, itself, stimulates the female into breeding condition.

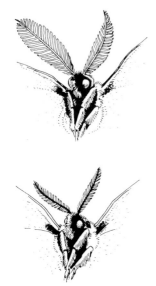

The difference in size of the antennae of male and female moths indicates the importance of a sense of smell in bringing the sexes together. The female gives off a perfume which guides the male to her, the molecules of the perfume being trapped on the large feather-like antennae of the male.

In the majority of animals, a female allows mating to take place only when she is in breeding condition. She is then said to be **receptive**. At other times the male is rejected, often aggressively. As the breeding season advances the female's reproductive organs develop and the ova ripen. At the same time her behaviour changes, she begins to accept the company of males and may even seek them out. Part of her new behaviour consists of indicating, by means of auditory, visual or olfactory signals, that she is ready to mate, that her ova have ripened. In mammals there is a cycle of changes, annual in deer and seals, every few months for dogs and every few weeks in rodents. This is the **oestrus cycle**. When not receptive, the female is said to be **in anoestrus**. When ready to mate she is **in oestrus** or 'on heat'. The oestrus cycle is repeated until interrupted by fertilization and pregnancy. Among rabbits there is no oestrus cycle. The female will mate at any time and the discharge of ova, or **ovulation**, takes place after mating.

Choosing a mate From a casual appraisal of the pairing of animals there is every appearance that it is the male, by his courtship behaviour, that is choosing his mate but the reverse is found to be the case. The male merely advertises to the world in general his willingness to mate. Among birds, males advertise their presence by songs and displays from the security of their territories. Each female makes a tour of inspection and accepts the advances of the male of her choice. Among mammals, the bitch on heat, for instance, attracts the neighbourhood's dogs but it is up to her to accept or reject them although in many instances her choice is random. The male's courtship is so indiscriminate that he does not necessarily confine his attentions to his own species. It is the female's negative action which normally prevents interbreeding. Mixed mating does sometimes occur, however, but in exceptional circumstances, as in captivity, where a rook will mate with a crow or a lion with a tiger. Occasionally interbreeding occurs in the wild, as, for instance, in Scotland where capercaillies are expanding their range. The hens move into a new area before the cocks and, while waiting, mate with Black grouse.

Where each individual has several mates, we speak of **polygamy**, from the Greek meaning much married. The term applies to either sex but one male having several mates is, strictly speaking, **polygyny**, meaning many wives. The converse is **polyandry** where a female mates with several males. Species may have a social system which is either

Elephant midwife
When an elephant is about to give birth she retires into the bush. As a rule she is accompanied by one or two other females. After a while the adult elephants leave their hide-out but now the formerly expectant mother is accompanied by her new-born baby, able to walk almost from the moment of birth.

This has been seen many times by explorers and naturalists, but the puzzle remained about what happened while the elephants were out of sight. Do the other female elephants accompany the mother-to-be perhaps to stand guard at a time when she would be unable to protect herself?

Twenty years ago Commandant Lefevre was in charge of an elephant training station at Gangala na Bodio, in what was then the Belgian Congo (now Zaire). He had the good fortune to be able to pry into this most private of occasions. For two hours he was able to watch all that happened. The birth was difficult and Lefevre saw one cow elephant use her trunk to deliver the other of her baby.

Such intelligent behaviour need not surprise anyone who has seen an elephant's brain. This is 8-9 ins across — not very large by comparison with the size of the elephant's body; but its surface is heavily convoluted.

Mad March hares
The hare family is widespread over the world and many of its members seem to go crazy in the Spring. In Europe it is the Brown hare that does this but the Jack rabbits of North America, which are near relatives, are tarred with the same brush. In Britain the saying 'Mad March Hare' has been used for centuries, because during the month of March

130

the males gather to chase each other back and forth, rear on their hind legs to box with the forepaws, or deal out tremendous blows with their hindfeet. The same goes for Jack rabbits. It has long been assumed this was because the males were fighting over the females, or that it was part of a courtship. The puzzle was that hares breed all the year round but go mad only in March. A recent investigation suggests that neither of these explanations is correct. To start with, only 16% of the year's conceptions occur in March, against 45% during the weeks after the madness has died down. Rather, it seems, the males by their fighting are establishing a breeding hierarchy, like that among male baboons.

Young male baboons mate with females at the onset of oestrus, but it is the dominant males that mate with them later and are responsible for most of the pregnancies. In this way the young males gain experience but the older males become the fathers. Perhaps hares have similar methods.

polygynous or polyandrous, or both, when it is known as polygamous. A strict one to one pairing is known as **monogamy**. In our own monogamous social system we are used to the idea of each person in a marriage making a very definite selection of partner, and mating is normally delayed until both partners feel certain that their choice is the correct one. By contrast we tend to think of 'brute animals' as neglecting such finer considerations although there is no reason why this should be assumed. It is true that even in some mammals mating is perfunctory, with no preliminaries, but where there is a long courtship it suggests that an element of choice is involved.

An example of a very long engagement is seen in the Wandering albatross where courtship is spread over several years. When a young male albatross has established himself on a patch of ground where he will eventually build a nest, he advertises his presence. Arching his long, slender wings, he points his bill at the sky and emits a loud trumpeting whistle. Unattached females glide overhead observing the males, each swooping low over several and eventually landing beside one. She approaches the male cautiously and the preliminaries of courtship are started. Both birds are wary of each other because, although both are sexually motivated, they both fear each other and the male may suddenly attack. **Courtship displays** and rituals are designed to overcome this fear of close contact. The female albatross approaches her chosen mate in a **submissive posture** designed to allay his aggression. In the gulls and skuas nesting around the albatrosses' colony, the female takes this further and mimics the action of a chick begging for food.

Initially, the courtship of the albatrosses does not proceed far and the female wanders off to visit other males. As months pass, she will spend an increasing length of time with each male and courtship gradually progresses. Eventually she stays with one male. Her choice has crystallized and courtship proceeds to culmination. The pair have formed a partnership that is dissolved only by the death of one member. The courtship displays, involving mutual nibbling of each other's necks, are touching to a human observer. It is hard to believe this has been a case of cold-blooded, purely random choice of mate. It is, however, difficult to prove it is otherwise. Deliberate choice of mate has, nevertheless, been demonstrated even in species where the pair bond lasts only for as long as is necessary for the act of mating and fertilization to take place successfully.

131

There are several families of birds in which the male contributes nothing to the rearing of the family. All his energies are diverted into displaying and competing with his fellows for the attentions of the female. To this end, the male develops a showy plumage, as in the peacock, the pheasant and the Birds of paradise. The males gather in special areas known variously as **leks, booming grounds, strutting grounds, arenas** or **hills**. Here, each male defends a small territory where he postures and calls. The ruff, a species of sandpiper, is a typical example. It is so named for the magnificent ruff of feathers which the male develops in the breeding season. In the spring the ruffs gather on a 'hill' where each stakes out an area about two feet across. They strut to and fro, with ruffs expanded and wings quivering, and the females, or reeves, are attracted to the hill where they make a tour of inspection. Then they make their selection and it transpires that each reeve chooses her mate from among the dominant ruffs. How she does this is not known but when several reeves visit a hill at once, they queue

Ruffs gathered on a 'hill' in spring are in an obvious state of excitement, each displaying his ruff at the slightest movement from one of his neighbours. From time to time a female, or reeve, recognizable from having no ruff of feathers, wanders in and selects one of the males as her mate.

for these males and ignore the others. The same happens with Prairie chickens, members of the grouse family. At one booming ground, there were nine males holding territories but one of these took part in 70% of all matings.

What is not known is how some males become dominant over others, nor how the females recognize them as such. At the strutting grounds of Sage grouse, young cocks hold territories around the periphery and move to the centre as they mature and old birds die off. So a hierarchy is formed and the females merely fly to the centre where they automatically find the dominant cocks. By contrast, a reeve walks around the hill where she could stop with any ruff. Yet she chooses a dominant bird even though his position is not indicated by topography.

In some mammals that live in social groups, the female may have little choice of partner because the males form a **dominance hierarchy** amongst themselves in which only the superior males may mate. The physical harrying of subordinate males by the dominant males may prevent them coming into breeding condition. On the other hand, it has been found that a bitch may show a marked preference for a certain dog, a fact well known to breeders of dogs and other animals, where the prize stud may be rejected in favour of the mongrel down the road. In an experiment in which beagle bitches were allowed to mate with any one from among a number of males, it was found that each would preferably accept one male more than any of the others. For instance, one bitch accepted a particular male in nine out of ten chances to mate. Each of the others had different preferences. Outside the heat period, the bitches still maintained preferences for associating with certain dogs, but the preferences came in a different order. It seems that a bitch may have an affinity for a particular dog as a companion during everyday contact, but when on heat the change in the dogs' behaviour towards her leads some to stimulate her sexually more than others.

A few animals, such as otters, may breed at any time of the year but each female has only a limited period of oestrus. In humans, however, the female is continuously receptive and this is seen as an adaptation to forming a permanent pair bond. Many birds — albatrosses, penguins, geese, swans and eagles among others — form permanent pairs. Familiarity speeds the animal courtship process, an advantage in time saving when the male has to help gather food and protect the young. Other primates form only temporary pair bonds during the oestrus period, but the male human is needed to help rear the very slowly maturing offspring.

Marriage, therefore, is not just a matter of personal preference comparable to what takes place in dogs or in Prairie chickens, where the females select the dominant male. The ability of a man to provide for the wellbeing of his family has to be taken into consideration. This practical principle seems also to have been demonstrated in terns, where an essential part of courtship consists of the male presenting fish to the female, a ritualized piece of behaviour which serves a useful purpose.

Courtship feeding, as this is called, is common among terns and also their relatives the gulls and skuas. When the pair has been formed, the male brings food to the territory and feeds the female. He continues to do this throughout the

One of the features of the courtship of the Sandwich tern is that the male presents his prospective mate with a fish. He continues to feed her throughout the courtship and afterwards when she is sitting on the eggs. Courtship feeding is not merely symbolic, but an indication that the male is capable of obtaining food — and the size of the eggs and survival of the chicks are dependent on his ability to do so.

Make no mistake about it, biologically speaking the male is less valuable than the female. She is the custodian of the next generation. His essential contribution is brief. So he is expendable. This much is emphasized by the behaviour of those species in which a courting male runs a high risk of being attacked or eaten by his intended spouse. The classical example is the spider, but it is not the only one.

Among insects the male Praying mantis stands a fair chance of being devoured while in the act of mating. A similar tendency is discernible in the Dance flies of the family Empididae, but the male empid fly averts disaster by an unusual and little understood stratagem. They are called Dance flies from their habit, similar to that of some midges, of gathering in swarms and flying to and fro. Both sexes may do this and the 'dancing' is associated with breeding. At the appropriate time a male enters the female swarm but not before he has caught one of the small insects preyed upon by empids. This he presents to a female, who plunges her piercing proboscis into it and starts to suck the life out of it. The male meanwhile mates with her after which she drops the prey.

An American empid goes further. The male wraps the gift insect in a frothy balloon before presenting it to the female, while in related species of Empididae the male merely spins an envelope of silk from glands on its front feet, which seems to placate the female equally well.

incubation period and only stops when the eggs hatch and he turns to feeding the chicks. The value of courtship feeding lies partly in providing extra food to the female to help build up the eggs forming inside her and partly in cementing the **pair bond**. In terns, it has been taken a stage further and courtship starts with the ritual presentation of a fish to the female. The female Common tern rarely feeds herself during the egg-laying period and is dependent on fish and shrimps brought in by her mate. The amount of food brought determines the size of the eggs, particularly the third and last of the clutch, and survival of the chicks is partly dependent, in turn, on the size of the egg. The male tern is also responsible for most of the feeding of the chicks when they are very young. Thus, it is to the female's advantage to find a mate who is successful at fishing. The ritual presentation of fish enables her to assess a male's potential as a provider and it seems female terns may leave a male if he does not come up to standard.

Reproductive strategy From this survey of courtship habits, a major theme is emerging. The aim of every male animal is to find one or several females with which to mate. It can be said that the whole point in life, at least in biological terms, is to leave as many descendants as possible and, according to Darwin's theory of evolution by natural selection, the best and most vigorous animals beget the most offspring. In other words, the survival of the fittest is only half the story; to be of use to the species the fittest individuals must breed well and pass on the characteristics that made them so fit to the next generation. The methods employed by a species to ensure this happening are called the **reproductive strategy**. As far as females are concerned, this means laying as many eggs or bearing as many young as possible, and for the male it means ensuring that he fathers the maximum progeny.

There are two basic ways by which this biological end is gained. A male may either mate with as many females as possible or he may ensure exclusive access to one or more females. The former is probably the simpler method and is effective if a female mates only once. Then the male is assured that her progeny is his own. This is the case with the Mediterranean fruit fly and it has enabled this pest to be eradicated on fruit farms. Males are attracted to a trap baited with an artificial version of the female's attracting pheromone. They are sterilized and released. Each sterile male mates with several females which, because they do not mate a second time, will never be able to lay fertile eggs. In other species, the female can mate several times so, unless males are scarce, there is the likelihood of a male's sperms being diluted by those from subsequent matings. Hence, we find males evolving methods of denying access by other males to their particular females.

Guarding a mate We have already seen some of these methods of denying access at work. One of the most common is the holding of a territory, the subject of the previous chapter. The males of a wide variety of insects, fishes, reptiles, birds and mammals are unable to obtain a mate until they have established a territory. Young, unestablished males wait in the wings, as it were, until the death of a territory holder allows them onto the stage of the breeding ground. The ability to defend a territory against competition is proof of the fitness of an individual to father offspring. Others maintain a hierarchy, which may also entail holding a territory, as in the Sage grouse described above. The workings of dominance hierarchies are discussed in Chapter 10.

Both territories and dominance hierarchies exist only where there is some kind of stable social system. Where this does not exist, particularly among the lower animals, males have had to find other ways of ensuring that their chosen females cannot mate with other males before the eggs are fertilized. One method is to guard the female, fighting off other males that approach. Another is for the male to remain clasped to her. Among butterflies and moths, for instance, copulation lasts for a long time. A week's copulation has been recorded for the Orange tip butterfly and the male Apollo butterfly becomes cemented to the female by secretions from his reproductive glands. During this time the eggs will be safely fertilized without the possibility of intervention by another male. The 'mating-tie' of dogs and foxes is thought to have the same function. The base of the penis becomes swollen during copulation so that the pair cannot disengage for some 15-20 minutes, perhaps as long as an hour. During this time, the dog dismounts so that he then is held back-to-back with the bitch. Although bitches show preferences, as we have seen, they will still accept several dogs in turn so the mating-tie will give the first dog's sperm a good start in the race up the reproductive tract to fertilize the waiting ova.

For frogs and toads, the guard phase has to come before copulation because the male sheds his sperm just after the

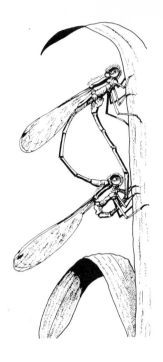

Tandem mating of damselflies, relatives of the better-known dragonflies. Male and female remain in conjunction until the eggs are laid. In this way the male ensures that the fertilization of the female will be by his sperms only.

Dragonfly mating
A common sight near a pond or stream in summer is two dragonflies flying in tandem. Even to the casual eye the insects are mating, but there is something odd about the posture adopted. There is, however, a simple explanation. This is a device to ensure that the female mates only once and then with a dominant male, the most fitted to sire her offspring.
Dragonflies show little in the way of courtship. The male overtakes the female in flight, lands on her back and grasps her with his legs. Then he curls his abdomen forward to grasp her

A harmless Grass snake's last line of defence is to sham dead. It rolls on its back with mouth open and tongue lolling. If it is turned the right way up, it immediately rolls on its back again and once more assumes the posture of death.

Left: Plaice and other flatfish change colour to fit their background. Within the skin is a mosaic of black, red and white pigment cells. Each cell can expand to show its colour more clearly or contract to reduce it to a dot. By altering the sizes of these cells the fish can match almost any colour or pattern within a matter of seconds. The camouflage is improved on sandy seabeds by the fish burying the edges of its fins in the sand.

Below: Praying mantis insects spend much of their time waiting, superbly camouflaged and motionless, for their prey to come within range. Their own enemies are also baffled. The Flower mantis, however, makes doubly sure. It can scare approaching birds by suddenly opening its wings to show the eyespots and by hissing at the same time.

A Shore crab regenerating a lost claw. Autotomy is the name for throwing off part of the body when it is seized by an enemy. Shore crabs can shed their legs or claws and grow replacements. There is a special line of weakness indicated by a groove where the hard casing of the limb snaps easily and the blood-vessels and nerves pass through a membrane at that point which seals rapidly so preventing bleeding. Breakage is achieved by a special muscle which contracts by reflex action to put an enormous strain on the breakage point.

Left: Fiddler crabs stake out territories on tropical beaches and display to each other with their enormous brightly coloured claws. Each species of Fiddler crab waves its claw in a particular way which is recognized as an invitation to mate by females and as a warning to rival males. The smaller male here has been slow in retreating from its larger rival and is being chased away after a brief fight with interlocked claws.

Below: When nesting in colonies, birds do not maintain large territories but defend just enough space to contain the nest and keep neighbours at beaks' length. Each sitting bird is just out of reach of surrounding birds, and this Greater flamingo is reaching up to nip another that has come too near.

The Jack Dempsey is a small freshwater fish from the Amazon. It is popular with aquarists but it is pugnacious, so it was named after the famous boxer. When the male comes into breeding condition he establishes a territory and if an intruder appears the two lie side-by-side, head to tail, with fins raised and colours heightened. If this show does not force one of them to retreat, they swim slowly in circles and attempt to butt head on or grab each other by the mouth and wrestle. As so often happens, the same displays are used in both territorial aggression and courtship but the female placates the male by keeping her fins lowered.

Left: The showy delicate Banded coral shrimp presents a prospective mate with a present of food. If she is already feeding the female may ignore the gift but if she accepts it the pair may mate for life.

Below: Courtship in birds starts with the male advertising his presence to passing females. The Wandering albatross spreads his 11-foot wings, points his bill at the sky and trumpets his availability. Interested females visit him and he will court many before finally one is accepted. This takes some time and Wandering albatrosses do not breed until they are about 10 years old. However, once a partnership has been formed, the pair stays faithful until one of them dies. A close attachment is needed because they take over a year to rear one chick.

Roman or Edible snails are hermaphrodite; each snail functions as both male and female. When they mate, both are acting as males and inseminate each other. Some time later in the summer, the snails become female and the sperms which have been in store since mating fertilize the eggs.

Reproduction without sex: a wingless Green aphid gives birth to live young without having been fertilized. She is known as the 'stem mother'; all her offspring are also wingless and female but some of their offspring have wings and migrate to other plants where they continue the process of virgin birth all summer. Only in autumn do males put in an appearance. They have wings and they mate with a new generation of winged females. The eggs from these unions lie dormant over winter and hatch out as 'stem mothers' in the following spring.

head with a special pair of claspers at the tip of his long abdomen. After this the female curls her abdomen and appears to be holding her mate by the front of his abdomen.

If we were privileged to catch sight of a fleeting action we should see the clue. Just before the male catches the female he brings his abdomen forward to transfer his sperms from his genital organs to a special accessory organ at the front end of the abdomen. This is why the female is pressing the tip of her abdomen onto that point. By doing so she takes up the sperms to fertilize her eggs. The fertilization is not instantaneous, and the two must remain in this position long enough for it to be consummated.

In most frogs fertilization is external, the female shedding her eggs into the water, the male fertilizing them as they are extruded. Maximum success is ensured by the male clasping the female in a close embrace known as amplexus.

eggs are laid so fertilization is immediate. A male will mate with several females if given the chance, but having found an unattached female he lays claim to her by clasping her body tight with his front legs, aided by special swellings which develop on the hands at this time. Such pairs are said to be **in amplexus**. The male discourages other males by croaking and thrusting at them with his hindlegs, so from his commanding position he can be sure that the ensuing spawn will carry his characters.

Guarding by proxy, in a manner reminiscent of that used by the crusader knights, involves the insertion of a plug in the female's reproductive tract. This occurs in some insects such as mosquitoes and the honeybee and in a few snakes. The American garter snakes and Water snakes, and perhaps others that have yet to be investigated, form a plug in the female's cloaca from a material secreted in the male's kidneys. It is inserted after ejaculation and hardens to block the oviduct. Any further matings are rendered ineffective and the plug drops away after a few days, by which time the ova will have been fertilized.

Sexual roles Almost by definition the male is the more active. At the most fundamental level the male's sperms are active in seeking out the immobile female egg. So is it often with the animal itself; the male seeks out the female and actively courts her. In extreme cases, the female animal may never move. The Gypsy moth is flightless and female bagworms, moths that are often a danger to fruit trees through stripping their leaves, do not leave their pupal shells. They are reduced to the status of egg-laying machines. Their legs and wings are reduced or missing and they do not feed. Males are attracted by a pheromone and mating takes place through the pupal case. The males, too, are reduced to travelling fertilizing machines. They are strong fliers but do not feed and are short-lived.

Other females are more active. Female albatrosses, ruffs and Sage grouse have actively to seek out the males who then start to court them, but the tables are turned with some birds where the females take the initiative in courtship and then leave the male to rear the family. **Reversal in sexual roles** is seen in the phalaropes, small wading birds of northern climes. The female phalarope courts one male alone and, after laying her eggs, she leaves him solely in charge. The habit is also found in the buttonquail, the dotterel, the Painted snipe and occasionally in the Bronze-winged jacana or lily-trotter.

145

The hen dotterel chases the male of her choice, separating him from the other males and inducing him to mate with her. Some hens lay two clutches of eggs in separate nests which are then incubated by two separate males. Later the hen may help tend the chicks. Little is known about the breeding habits of jacanas and Painted snipe except that the females hold territories, but the buttonquail female attracts a male with a booming call and then courts him, circling with tail raised and chest puffed out and stamping, while she continues to boom. The female Barred bustardquail, an Asian buttonquail, even develops a bright courtship plumage during the breeding season. Naturally, the basic biological roles of egg-laying and insemination cannot be reversed but so 'masculine' are hen buttonquails that Indians use them for 'cock-fighting'.

Buttonquail female (right) in courtship display to the male.

Hen-pecked husbands

Throughout the animal kingdom, in species in which there are differences between male and female (sexual dimorphism), it is the rule that the male wears the most colourful costume, whether it be scales, feathers or fur. It is he that carries the personal adornments. It is also he that does the courting and usually he leaves the female to look after the family. The exceptions are so rare as to compel attention. The outstanding example among birds is the phalarope, known as the Grey phalarope in Britain, after its winter plumage, and as the Red phalarope in the United States after its summer plumage.

The female phalarope does the courting, the male builds the nest, incubates the eggs and tends the chicks. Sometimes he attracts the attention of two females and then he cowers submissively while they squabble over him.

The female emu of Australia also does the courting. She summons the male to her with a loud call. It is the male emu that makes what little of a nest they use and it is he who incubates the eggs and protects the chicks.

Even among rhinoceroses, the female makes all the advances and she has been seen to chase an unwilling male for miles.

The one outstanding exception to the male wearing all the finery is found in the human species, but not among all men. Among primitive tribes it is the males that adorn themselves the most and even in later stages of civilization the male is the more resplendently dressed. It is only in the modern so-called Western civilization that man has taken to wearing more drab apparel than the woman.

In the lower levels of the animal kingdom there may be a combination of sexual roles in one individual. Each animal has functional organs of both male and female and is known as a **hermaphrodite**, a name derived from the son of the Greek gods Hermes and Aphrodite. This condition is also found in plants where male stamens may be found with the female stigma in one flower. Hermaphroditism in animals is generally found in those that are sedentary or at least slow moving. There is usually a mechanism to prevent self-fertilization taking place. Earthworms are functionally both male and female at the same time and during copulation a pair of worms fertilize each other. Mating can sometimes be seen on warm, damp nights because some species of earthworm come to the surface to pair. Two worms stretch out of their burrows and lie side by side, heads pointing in opposite directions, for two or three hours. Their bodies are held together both by a sticky substance secreted by the clitellum or 'saddle' and by the chaetae or bristles. Sperms are passed from the male organs along a groove running down the body and into the female organs of the other worm. Here they are stored in a spermatheca until egg-laying. Copulation lasts two or three hours.

Snails have a more elaborate form of hermaphroditic mating in which the pair stimulate each other with surprising violence. There is only one gamete-producing organ, called the ovotestis. At the beginning of the breeding season, all snails are male and the ovotestis produces sperms. One snail finds another by following its slime trail and the two circle each other in decreasing circles. On drawing near, each projects a 'love dart' into the body of the other, which penetrates the flesh to a considerable depth, acting as a massive stimulus to mate. After mutual insemination, the sperms are stored in a spermatheca, a chamber leading off the oviduct, until, in midsummer, the ovotestis produces eggs.

Sex reversal is a form of hermaphroditism which occurs in a number of animals for no very clear reason. Sex is determined genetically and sexual characters develop through the action of hormones yet female Common frogs change sex if kept at 26.6°C (80°F) and the amphipod crustacean *Gammarus duebeni* produces female offspring at high temperatures and males at low temperatures.

The Slipper limpet, which is more closely related to periwinkles than to Common limpets, was accidentally introduced to European waters from North America. Slipper limpets live in piles of eight or nine individuals fixed on one another's backs. They have become a pest to oyster beds as they compete with oysters for living space and planktonic food. Our interest in them is that they change sex during their lives. The first Slipper limpet to settle automatically becomes a female. The next to settle on her starts as a male and becomes a female as it grows older, so that a chain of Slipper limpets consists of females at the bottom and males at the top with individuals of intermediate sex in the middle.

Slipper Limpet

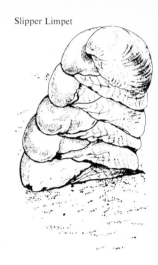

Chapter Seven
Raising the Family

larval life — ovovivipary: ovipary: vivipary — parental care —
insects: social insects — fish and parental care: mouthbrooders —
parental care in amphibians and reptiles — parental care in birds:
brooding: incubation: altricial and precocial young: nest building:
fledging: brood parasites — ground nesting: cryptic coloration: egg
rolling: nest hygiene: imprinting: recognition of eggs and offspring:
teaching young — mammal parental care: infancy: weaning:
retrieving and carrying: teaching young: primates: play

The result of an act of mating is a fertilized egg. This not only contains the germ of a new individual, but is furnished with a food store that supplies energy for development, and eventually a young animal emerges. The young animal appears in a variety of forms. It may be a **larva,** a free-living animal, quite different in appearance from the adult. The most familiar examples are caterpillars and tadpoles, which are the larvae of moths and butterflies and of amphibians, respectively. Many marine animals, such as crabs, barnacles, Sea anemones, Sea urchins and starfish produce minute larvae that float in the sea as plankton before settling on the sea bed and transforming into adults. These marine animals are sedentary or slow-moving and the free-floating larvae serve to distribute the species. The reverse is true for caterpillars; they do not move far but the adults can fly, and it is they that keep the species spread out. In other animals the larval stage is by-passed and the egg develops directly into a miniature adult. Snails, spiders and birds are examples of this.

Among those animals whose offspring develop directly without a larval stage, the fertilized eggs may be released into

It is often said, erroneously, that the mother otter teaches her young ones to swim. Although otters are fully adapted to an aquatic life, the young ones often need to be encouraged to enter the water to take their first swim, the mother enticing them in by showing them food.

150

the water or laid on land and left to their fate. Vast numbers of the offspring die and, consequently, vast numbers of eggs have to be laid. This is wasteful and many animals have taken to laying fewer eggs and assisting their survival by giving them some means of protection. At the simplest level, the mother hides the eggs in a crevice or a hole in the ground, or she carries them attached to her body. Also, each of the smaller number of eggs gets a larger share of the food store supplied by the mother and development can proceed further before hatching. This is the condition that leads to the direct hatching of miniature adults. Another form of maternal protection is to retain the eggs within the body. In some cases the eggs lie in the maternal reproductive tract, which acts as no more than a living nest or incubator until the eggs hatch. This is called **ovovivipary**. It is a stage between egg-laying (or **ovipary**) and the bringing forth of live young (or **vivipary**). In true viviparous animals there is a connection between mother and offspring through which food and oxygen pass to the developing embryo and carbon dioxide and excretory products are taken from it. Vivipary is extremely important in mammals because the embryo has to be kept warm throughout its development. Birds, which are also warm-blooded, lay eggs as a means of saving the weight carried during flight. They keep the embryo warm by incubation. True vivipary has developed in a number of other animal groups. Among the fishes, there is a gradation from retaining the eggs in the oviduct for a while, through supplementing the yolk with nutritious secretions from the maternal tissues, to full vivipary, as in some sharks, where there is a direct link between the oviduct wall and the yolksac. In the reptiles, too, there is a gradation between ovipary and vivipary. Most reptiles lay eggs but some snakes and lizards retain their eggs until hatching (ovovivipary). The Common or Viviparous lizard of Europe is one such and, even then, it also lays eggs in some parts of its range. Others, such as some skinks and the European adder, are completely viviparous. The yolk is supplemented by food passed from the maternal tissues. One advantage of vivipary in reptiles is that the embryos benefit by the mother basking in the sun or retreating underground in cold weather, which explains the ability of the adder to live beyond the Arctic circle.

In this chapter we are principally concerned with the care of offspring after they have left the mother's body. Of the hundreds of thousands of species in the animal kingdom

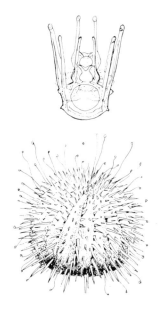

A larva is the young of an animal, but it is something more: it represents a stage in the life history at which the individual has a markedly different shape to that it will have when fully grown, as in the larval (above) and adult (below) Sea urchins seen here.

only 0.1% show any indication of parental care. Most of the aquatic species shed their eggs and sperm into water and make no further contribution to their welfare. The remainder, in which fertilization is internal, lay eggs and the responsibility of the parents ceases forthwith. Nevertheless, there is a kind of parental care shown in a minority of most animal groups. In the birds and mammals, where parental care reaches its peak, there is a heavy responsibility, particularly for the female parent.

Insects

Parental care is well developed in some insects. Again, the majority show no parental care but the ants, bees and wasps of the order Hymenoptera and the termites (the so-called white ants) of the order Isoptera are often referred to as 'the social insects'. This is not to say that all species in these four kinds of insects live in colonies, or societies; in fact, the majority of them are solitary. In the solitary species the most that happens is that the female makes a shelter for the eggs which she provisions with food for when the grubs hatch out. But in a fair-sized minority the adults live in societies consisting of a queen and numerous sterile females, or workers. The workers care for the eggs and tend and feed the larvae until they have become adult.

Earwigs show some parental care; the female stays with her eggs until after they have hatched. Moreover, if the eggs happen to be moved, she will gather them together again and continue to brood them. The important point is that one generation makes close contact with the next and so the scene is set for the advancement in parental care to that shown by the social insects, as well as the birds and mammals, especially man, where the parents can teach the young their acquired skills.

The progress from a situation in which parents merely meet their offspring by accident to the development of a society consisting of several generations is best illustrated by the different species of **bees.** The majority of bees, as we have seen, are solitary, that is they lay their eggs separately and abandon them but in the genus *Halictus* we see the beginnings of a social life. There are thirty-five species of *Halictus* bees; in these each female digs a deep burrow in the soil with a nest at the bottom where fifteen or so eggs are laid. Each egg is placed in a specially constructed brood-cell which is provisioned with pollen for the bee larva to feed on. Collecting a store of food for the next generation could be

Cuckoo bees

Brood parasitism is not confined to the birds. Cuckoo bees of the genus *Psithyrus* look very similar to bumblebees, but close examination shows that they lack the pollen baskets on the hindlegs. They have no need to collect pollen because they make no nest and have no brood cells to provision. Instead, the female Cuckoo bee enters the nest of a hard-working colony of bumblebees and, after a fight, she kills the resident queen and takes over. Thereafter the bumblebees have to rear Cuckoo bee queens and drones. It seems that Cuckoo bees arose from bumblebees, because it is not unknown for queens of one bumblebee to enter nests of other species and take them over.

Among the ants there are a number of parasites that enter nests, kill the queens and take over; but one, *Strongylognathus testaceus* does not kill the resident queen. She is allowed to continue laying but all her offspring become workers destined to look after the growing brood of parasites, all of which become fertile males and females.

152

Young honeybee worker inspecting a grub in its cell. The first tasks of a worker after pupating include tending the next generation of grubs in the cells of the comb.

termed the beginning of family life, but some *Halictus* bees go one step further and stay with the brood, guarding them, in some species, until the young bees emerge.

The guarding stage represents a lengthening of the life-cycle because it is usual for insects to die after mating and egg-laying, their life's work having been completed. This extension of life span brings the two successive generations together and, in one species, *Halictus malachurus,* the female lives for the rest of the summer with her offspring. The first brood to emerge are all infertile females. They stay with their mother, taking over the duties of pollen collection and building brood chambers. The original bee now concentrates on egg-laying and in late August she produces fertile females and males. They mate and the young fertilized females survive the winter while the rest of the little colony dies off. *Halictus malachurus* represents the endpoint of this line of evolution but a simple social system with a female, a queen, reigning over a family of infertile females, the workers, is also found in bumblebees and it is developed to a greater degree in the honeybee and the wasps, ants and termites. Their intricate social organizations are discussed further in Chapters 10 and 11.

Fishes

Fishes, amphibians and reptiles are, on the whole, careless parents but there are notable exceptions. In *King Solomon's Ring,* Konrad Lorenz has described one of the most charming stories of animal behaviour. Lorenz had an aquarium stocked with a pair of jewelfish, the most beautiful of the cichlid fishes. The cichlid family is popular with aquarists and is unique among fishes in that the male and female stay together to rear their young. They are, in fact, married.

Each evening the baby jewelfish are brought back to the nest, a scoop in the sand where the eggs were laid. The female lies in the nest and the babies gather around her, while the male searches the neighbourhood for any wayward offspring. Any he finds are simply scooped up in his mouth and spat out into the nest. One evening, Lorenz was feeding the jewelfish with pieces of worm while the babies were being 'put to bed'. The male seized a piece of worm but, while chewing it, saw a straying baby. He immediately took it into his mouth with the worm, but stopped dead in his tracks. His mouth now contained a worm to be swallowed and a baby fish to be spat out — an obvious **conflict situation** with the

motivation of two antagonistic drives. For a moment there was stalemate but the conflict was resolved to the satisfaction of both feeding and parental urges in a manner that would have done credit to the wisdom of a human being. The jewelfish spat out both worm and baby, gobbled up the worm, then retrieved the baby and took it home.

The active parental role of the male jewelfish is paralleled by several other species. It seems that many male fish make better parents than do the females. The male Siamese fighting fish, renowned for its fierce battles in territorial disputes, is a model father. He constructs a floating nest, by taking gulps of air and blowing mucus-coated bubbles which stick together in a raft at the surface. As the female lays her eggs, the male takes each one in his mouth, covers it with saliva to make it buoyant and places it under the bubble nest. The female lays her hundred eggs in batches of about half a dozen so the male has time to collect them. After egg-laying, the male drives the female away and guards the eggs until they hatch a day or two later.

The male Three-spined stickleback is another good father. In contrast to the Fighting fish's floating bubble nest, the stickleback makes a nest, hidden amongst the weeds, from

Male Fighting fish of Thailand, constructing a raft of bubbles on the surface of the water. As the female lays her eggs he takes each in turn and places it on the underside of the bubble nest.

154

pieces of water plant glued together with a sticky secretion from his kidneys. The female is induced to enter the nest to lay her eggs, after which she departs and the male then enters to fertilize them with his **milt.** The eggs hatch in five to twelve days, depending on the temperature, and throughout this period the male stickleback stands by the nest fanning the eggs with his pectoral fins. This ensures that the eggs are constantly supplied with a stream of fresh, aerated water that probably helps to keep them clear of fungal infections. When the eggs hatch the male continues to guard the baby fishes and retrieves any that stray, but after two weeks they start to swim about in a school and his interest wanes.

Among the cichlid relatives of the jewelfish there is a variation of parental duties. The genus *Tilapia* consists of about a hundred species of fishes which are widely kept in aquaria or cultivated for food in fish farms around the warmer parts of the world. Some tilapias have a fairly straightforward family life. Male and female clean the surface of a stone and the eggs are glued to it. Both parents fan the eggs and the young fishes are shepherded by the female. However, most tilapias are **mouthbrooders.** The male, the female, or both, depending on the species, take the eggs into their mouths as soon as they are laid. The eggs hatch there and the young eventually emerge, although they dash back into the parental mouth when danger threatens. Inside the mouth, the eggs and young fishes are kept aerated by the flow of water passing to the gill chambers, and the constant movement against the sides of the mouth keeps them clean of fungi and bacteria.

Amphibians and reptiles

The amphibians and reptiles, as a rule, are poor parents. The eggs are laid and the parents depart. The spawn of amphibians eventually hatches into tadpoles which, while young, cling to water plants with a sticky secretion on the head. At first respiration is by external feathery gills, but these are later replaced by an internal set of gills, except in newts. Gradually the legs develop, the gills are replaced by lungs and the young amphibian is ready to come out on land. As with the fishes, if there is any parental care, it is the male that most frequently takes on these duties. The male South African bullfrog remains with the spawn and, being aggressive, protects it and, later, the tadpoles. The male Midwife toad gathers the spawn, which is laid in strings, around his hindlegs and retires under a stone, emerging

Crocodile paterfamilias
In recent years observation of the habits of the Nile crocodile have shown that both the mother and the father crocodile stay around the nest and show a marked degree of parental care.

When the young crocodiles hatch, the mother takes them one by one in her jaws and flips each in turn to the back of the mouth. Since a clutch of eggs may be anything from 16 to 80 or more, she has to depress her tongue until a pouch forms under her chin to accommodate them. Then she takes them to water, opens her mouth, and sways her head from side to side to wash the sand from the hatchlings.

The father crocodile may also take hatchlings to water to wash them. He has even been seen to take an egg into his mouth and roll it back and forth between his tongue and palate to crack it, in order to help the hatchling to leave the shell — a delicate task for jaws capable of cracking the long bones of large antelopes.

nightly to damp the eggs in water. South American frogs, such as the Barking frog, also guard the spawn under a stone but keep it moist with fluid from the cloaca. An exception is the Marsupial frog, of South America. As the eggs are laid the male pushes them with his hindlegs into a pouch on the female's back. She carries them until the froglet stage is reached.

As we have seen, some reptiles exhibit parental responsibilities in the form of ovovivipary, retaining the eggs in the mother's body until they hatch, and vivipary, giving birth to live young. Beyond that there is little sign of maternal care except among crocodiles. The Nile crocodile lays her eggs in a pit and, after covering them with sand and vegetable rubbish, stays with them guarding them from predators. Just before the young crocodiles hatch, they begin to grunt and their mother starts to uncover the nest. The babies stay with her for a few days, following her like ducklings, in search of insects and yapping if they get lost. Among snakes, parental responsibility is limited to some cobras and pit-vipers lying coiled on the nest to protect the eggs; a few snakes, also, are viviparous.

Breeding in blizzards

The prize for rearing a family under the most adverse conditions must go to the Emperor penguin. The breeding season begins in April to May, the start of the southern winter. Each female lays one large egg. No nest is made but both parents share in the incubation. At first they pass the egg from one to the other, holding it on their feet, covered with a fold of abdominal skin, to protect it, and later the chick, from the intense cold. After a few days the female leaves, waddling across the ice to open water to feed for the next two months.

During her absence the male must fast. Balancing the egg on his feet he shuffles about the rookery. During blizzards and very cold spells, when the temperature may drop to -80° C (-110° F), the males in a rookery huddle together to keep warm, so conserving heat which would otherwise have to be supplied by using up their blubber. As a penguin on the outside of the huddle gets cold he forces his way into the middle to warm up.

At the appropriate time the female returns fat with blubber and with her crop filled with fish for the chick, but not necessarily to her mate. She teams up with the first male she meets, relieves him of his egg and he goes off to open water to take his turn at gorging food to replenish his blubber. Should the females be late in returning, so that the eggs hatch before their arrival, the males feed the chicks with a secretion from their stomach lining.

As the chicks grow older they come out from under the parents and gather in nursery groups known as creches. When the temperature falls they huddle together to keep warm.

The billing and cooing of doves and pigeons is so familiar a phenomenon and so well known that we apply the term to the amorous antics of humans. In the birds it is not so much a display of affection as the start of a subtle series of reactions without which the species could not be perpetuated.

In doves and pigeons both parents share the parental duties. Both incubate, both feed the young on what is known as pigeon's milk. About four days before the eggs are due to hatch there develops in both birds a crop gland which secretes the 'milk', an enzymic fluid that pre-digests the food regurgitated to feed the newly-hatched squabs. Deprived of this 'milk' in its early days the squab (or chick) cannot assimilate solid food.

If in the course of incubation the eggs should become addled both parents will sit for 18 days but no crop gland develops. It is as if some quality in the egg itself governs their behaviour. Provided the eggs are viable all goes according to a set pattern and schedule, and this it seems is governed by the exchange of saliva between the two parents during billing and cooing. A male bird that has had no contact with the female can be made to undertake incubation and to develop a crop gland after being injected with the saliva of a sitting female.

Birds and mammals

Compared with other groups in the animal kingdom, parental care is very much the prerogative of birds and mammals. The young are fed, protected, kept clean and warm, even helped to learn to fend for themselves. While mammals have evolved live birth and the feeding of the young with milk produced in the mother's body, birds have retained the egg-laying habit of their reptilian ancestors.

Birds   The eggs are incubated in a nest by one or both parents. They are kept to within a few degrees of the adult body temperature by being balanced on the parent's feet and covered with a bare patch of skin, known as the **brood patch** which is well supplied with blood vessels and acts as a hot-water bottle. There are a few exceptions to this rule. King and Emperor penguins make no nest but carry their single egg on their feet. The Australian megapodes, such as the Mallee fowl, keep the eggs warm in the heat generated by the sun's rays, volcanic action or decomposing vegetation. Cormorants, gannets and boobies have no brood patch. They wrap their webbed feet around the eggs. Ducks also have no brood patch, but they pluck some of the breast feathers to make a nest lining.

Incubation behaviour is carried out by the female alone, by both sexes or, rarely, by the male alone. Where both sexes sit on the eggs, incubation stints are shared and there may be an elaborate take-over ceremony. A frequent case of breeding failure is one parent failing to return to the nest or leaving it before its mate returns. In most birds, incubation does not start until the entire clutch is laid. Consequently the eggs hatch together but in some birds, notably hummingbirds, birds of prey and parrots, incubation starts with the laying of the first egg and hatching is asynchronous. The chicks are of different sizes and the younger ones may be forced out of the nest or may fail to get food and perish.

Throughout incubation, the eggs are turned at intervals so that they are evenly warmed. Then, a few days before hatching, the chicks **'click'**, a sharp sound synchronized with their breathing. A little later they start to hatch. Using the chalky eggtooth on the tip of the bill they hammer a ring of small holes in the blunt end of the egg. This is known as **pipping** or **chipping**. Then, by straightening the body, the chick forces off the shell at the ring of pips and struggles out.

The newly hatched chick is wet but its down soon dries out and becomes fluffy. At first it has to be **brooded** by the parents, for a length of time depending on the species and, to

157

a certain extent, on the weather. The care of the chicks falls into two broad categories. Some stay in the nest until they can fly and are known as **altricial,** others leave the nest soon after hatching and are called **precocial.** Alternative terms are **nidicolous** for those staying in the nest and **nidifugous** for those leaving it. The advantage of using the first of these is that the words can also be used for mammals. So far as birds are concerned, it is usual to call altricial young **nestlings** and precocial young **chicks.**

Altricial birds   A main difference between altricial and precocial species is that the former build more elaborate nests. Their nests have to retain the growing and often quite active young. The typical, cup-shaped nest of hedgerow and garden birds of the Passeriformes (perching birds) are designed to hold altricial young. Nests are built, often by the female alone, in a set series of innate behaviour patterns, although experience plays a part in improving the manipulation of material and the selection of good nest sites. The basic method of forming the nest cup is to push alternately with the legs and swivel round, pressing and moulding with the breast so that what appears to be a woven structure has been worked in the manner of felt.

The young of altricial birds need feeding until they leave the nest. This is a full time chore for the parents who forage throughout the daylight hours. At first the chicks are blind and almost helpless. Apart from sleeping, all they can do is beg for food with a **gaping response.** Typically, the young nestlings respond to vibration, which signals the parent alighting on the side of the nest. They immediately rear up, stretching their necks and opening their beaks in a wide gape. The lining of the mouth is a bright colour, often yellow, and acts as a releaser, causing the parent to deposit food in the yawning mouth.

When satisfied, the chick lowers its head and goes back to sleep. After a few days the eyes open and the chicks now respond to the sight of a parent but the gape is not directed at the parent until the nestlings are ten days or so old, when they then direct their gaping at the parent's head. The parents feed the nestlings indiscriminately but as replete nestlings stop begging, all are fed in turn unless there is a shortage of food. Then, the weaker chicks get edged out and eventually die of starvation or are preyed upon by a weasel, a jay or other predator attracted by their continual clamour. As the nestlings grow, their coat of down is replaced by feathers that sprout from waxy sheaths. The development of

Home builder
Many birds build beautiful cup-shaped nests of interwoven material. Casual examination of the nest gives the impression that the bird has patiently threaded the strands of grass, horsehair or other materials, in the way a basket-maker might work.
There are few hour-by-hour accounts of how the nest progresses, but one tells of a blackbird that suddenly arrived one morning and placed two twigs on a windowsill. He did this several mornings running. Then his hen came with more twigs and placed these in the little heap which eventually grew to a sizeable pile of twigs and grass. The following morning at 6.30 a.m. the hen was seen hopping about on the pile of material, making a circular movement as she hopped round and round amongst the twigs and grass. By the evening the hen was sitting in the middle of a cup-shaped nest, transformed from an untidy heap of materials into a perfect nest.
While the cup was being formed the bird was using two types of movements: scrabbling by pressing down with the body and pushing back with each leg alternately; and turning, which is done in the sitting position. Finally, with the bill the bird tucks in the loose ends, pulls others into place and it is this that gives the illusion that the whole fabric has been meticulously woven.

Female House sparrow, seeing her fledgling in difficulty on its first long flight, takes off from her perch and flies underneath her offspring to give it suffic-ient lift to carry it to a safe landing.

the feathers is known as **fledging** and a young bird with a complete set of feathers and able to fly is a **fledgling.**

The gaping response is also used by **brood parasites** (or **nest parasites**), like the European cuckoo, that rear their young without effort by laying their eggs in the nests of other birds and leaving them to act as foster-parents. The female cuckoo watches for a small bird to finish its nest and start to lay. Then, when the bird is away, she slips in, removes an egg and substitutes one of her own. The cuckoo's egg often matches that of the specially selected host species but this does not happen in Britain where a variety of hosts are used.

The mimicking of the host eggs reduces the chances of the nest being deserted. If all goes well, the baby cuckoo hatches before the other eggs and then proceeds to eject them from the nest by manoeuvring them onto its back and hoisting them over the side. The cuckoo grows rapidly and leaves the nest after three weeks, although it continues to be fed by its foster-parents which, being so small by comparison, have to perch on its back to reach its mouth. The huge gape acts as a **supernormal stimulus** and induces other small birds also to feed it rather than their own chicks. Not uncommonly, when a baby cuckoo is calling for food, small birds on the way to their own nests with food for their own nestlings, will make a digression to push the food down the young cuckoo's throat.

Not all members of the cuckoo family are brood parasites. The Indian koel lays its eggs in the nests of crows and its young survive alongside the young crows by mimicking their appearance. On the other hand, the American members of the cuckoo family, such as the anis and the roadrunner, raise their families in the normal manner. Brood parasitism is employed by some of the cowbirds of the American troupial family. The Bay-winged cowbird, surprisingly, parasitizes the Screaming cowbird simply by ousting it from its nest before the eggs are laid. The Brown-headed cowbird lays its eggs in the nests of over two hundred and fifty species, with a preference for orioles and Song sparrows. The young cowbird does not eject its nest mates but generally outgrows them and wins in the competition for food. Some birds have a defence against cowbirds. Tyrant flycatchers build a false floor over the trespassing eggs and American robins throw them out of the nest. Other brood parasites are the African honeyguides, whose nestlings have hooked beaks for killing nest mates, and whydahs which parasitize Weaver finches.

Mimicking the eggs and chicks of the host is particularly important for a brood parasite because it is usual for a bird

159

to remove foreign objects from its nest. This is a trait found in both altricial and precocial birds, but is less developed in precocial birds where the nest is abandoned soon after hatching. With both kinds of birds, the empty eggshells are removed after the chicks have hatched. Not only may the shell smother or injure the chicks, its white interior may catch the eye of a predator. The parent bird recognizes that the shell is empty by the white membrane edging the broken rim and by its light weight. Altricial birds also remove faecal sacs, the membranous envelopes in which nestlings eject their droppings.

Precocial birds   The precocial birds are typically those that nest on the ground. Ground-nesting is dangerous because the nest is very vulnerable to attack from foxes, weasels and other ground predators. Vulnerability is reduced by the cryptic coloration of both sitting birds and eggs. Among many precocial birds, incubation is carried out by the females alone and we find that female ducks, pheasants and grouse are drably coloured in comparison with the males with their gaudy courtship plumages. Their best defence is to sit tight to avoid detection and to take flight only when danger is imminent. The unexpectedly explosive rising of a sitting bird may well be sufficient to distract the predator from finding the eggs.

A danger to individual eggs in the shallow, scoop nests of ground nesters is that of being knocked out of the nest, but, as we have seen in Chapter 1, birds have a fixed **egg rolling** behaviour in which they retrieve the egg by scooping it back into the nest with the bill.

Precocial chicks are very active soon after birth and they set out with their parents on foraging expeditions. Many can feed themselves immediately but others such as the young of coots and moorhens have to have food brought to them. The chicks of barnyard chickens instinctively peck at small objects that contrast with the background. In a normal environment this means fallen seeds, insects or small pebbles. Over the course of a few days the chicks' aim improves, through development of the nervous system (i.e. maturation) and through practice. They also learn what is edible, rejecting pebbles and swallowing seeds and insects.

Imprinting   It is essential for the survival of a precocial chick that it stays with its parents. If it becomes separated it is doomed. Therefore, a close link between parent and chick must be formed by the time that it leaves the nest. The chick learns the identity of its parent by the process of **imprinting.**

The eggs of ground-nesting birds are sometimes pushed out of the nest accidentally. The parent retrieves them. This oystercatcher, with a larger than life-size egg placed near her nest, chooses to retrieve it in preference to her own egg. The larger egg acts as a super-normal stimulus.

Oystercatchers teach their chicks to feed

Oystercatchers are large shore birds with pied plumage, a long red bill and red legs. They feed on limpets, mussels, cockles, winkles, crabs and worms, but not all oystercatchers take the same food, and investigation into their feeding habits revealed an unexpected linkage between the food taken and the division into communities. Indeed, the differences in feeding habits are so marked that populations of oystercatchers are distinctly divided by them. Moreover, mussel-eaters mate only with mussel-eaters, cockle-eaters only with cockle-eaters, crab-eaters only with crab-eaters. The distinction amounts almost to a gastronomic culture, perpetuated by the way young oystercatchers learn from their parents which foods to take.

To begin with, the oystercatcher chicks peck only at empty shells on which the parents have fed, picking up the pieces of flesh left in them. Later, they take shellfish which the parents have opened and remove the whole flesh by themselves. During this graduation they are learning the actions of the bill needed for dealing with a particular prey. Finally, they are able to carry the whole feeding process through for themselves, serving an apprenticeship by opening only small shellfish, graduating to larger ones as they become more proficient.

Throughout, the investigators found a rigid adherence to one kind of food. They never saw mussel-eating chicks eat crabs. In fact they were afraid of them. And a crab-eating chick totally ignored cockles and mussels.

This is a very rapid, non-reversible and permanent form of learning that only operates within a very short period of the early life of an individual, known as the **'sensitive period'**.

The first object of a suitable size that a chick sees after it hatches is indelibly stamped on its mind as its parent and it becomes the focus of a **following response.** The chick follows it blindly through all obstacles. In nature the object of imprinting will invariably be its parent but under experimental conditions young chicks can be imprinted on a variety of bizarre objects such as toy animals and wooden boxes. The phenomenon was first studied by Konrad Lorenz who, as described in *King Solomon's Ring,* induced broods of goslings and ducklings to imprint on him. He found that the goslings accepted anything as their mother but that the mallard ducklings required a foster-parent who was about the right size and made the correct noises. So, to induce ducklings to follow him, Lorenz had to creep about on his knees, quacking continuously, to the amazement of his neighbours.

For ducklings, then, an important stimulus for imprinting is the continuous low quacking of the mother or her human substitute. The imprinting on her voice starts before hatching, just after pipping, when she and the chicks call to each other. Thereafter her voice is a beacon and reassurance to the chicks, and their frantic voices alert her if they become separated. After hatching, the feel of the mother's body is a stimulus to the duckling. This contact keeps the chick quiet and still and, thereafter, whenever the mother gathers her brood under her, as in times of danger, the ducklings lie quiet to avoid detection.

Imprinting is also featured by those mammals which have active young, such as young hoofed animals which must keep up with their mothers within a few hours of birth. Again, human beings are accepted as substitutes and foundling lambs become imprinted on the farmer's wife who feeds them. So specific is this imprinting that she can later leave the lamb outside without danger of its straying by hanging her apron on the clothes line. If frightened, the lamb will rush to the apron for safety.

Recognition of offspring The degree to which birds and mammals recognize their offspring varies from species to species and depends on circumstances. The Grey-headed albatross, for instance, recognizes the position of its nest and will feed any nestling placed in it. This is sufficient because young Grey-headed albatrosses do not leave the nest until

161

they take to the air, at which point parental duties are finished. On the other hand, young Wandering albatrosses leave the nest and waddle about exercising their wings in preparation for flight. Consequently, parent Wandering albatrosses have to search for and recognize their offspring. Guillemots nesting on cliff ledges have to recognize both egg and chick. No proper nest is made and the egg is recognized by the pattern on the shell. Chick and parents learn to recognize each other before hatching by calling to each other, and when the chick eventually leaves the ledge and flutters down to the sea it is joined by one parent which has recognized a special call.

Among mammals smell is very important as a means of identifying the offspring. A ewe will accept any lamb presented within eight hours of giving birth but if she has been able to lick her lamb for over twenty minutes she will reject all others, unless she is deceived by their being rubbed with the skin of her own lamb or with birth fluids. Later, she will recognize her own lamb's voice or appearance, although she will always confirm the identification by smell. Seals employ the same system. As soon as her pup is born, a Grey seal cow turns to smell it, a process repeated several times

Fastidious hog
A party of tourists was assembled in a hide, in the Infolosi Park in southern Africa. They were watching a warthog sow and her two piglets wallowing. Finally, the sow emerged from the wallow completely covered in mud, followed by her two piglets, one of which was well caked all over with mud and the other of which was muddy except on the back. The trio had not proceeded far when the sow looked at her offspring with a clean back and then nudged it towards the wallow and drove it in again. The sow also entered the wallow and rolled in the mud once again, but on her back, as if trying to indicate to the piglet that this was what it should do. After a while the piglet appeared to have got the message, turned on its back and wallowed, whereupon the sow got to her feet and walked out of the wallow with her youngster following her.
The tourists were highly amused and gave vent to a chorus of laughter.
People often talk glibly about an animal teaching its young, but concrete examples of this are hard to come by, so much so that some ethologists declare that an animal never actively teaches its young, but merely educates it by example. There may be some truth in this, but to the assembled tourists the scene they witnessed was a clear-cut case of animal education.

The mother hippopotamus teaches her offspring that in the water it must swim level with her shoulder, on land it must follow at heel: these are the best positions for defence of her offspring.

It was a warm still day with a persistent drizzle as the Grey squirrel set to work independently and most industriously to build a drey in the fork of a large oak. All the twigs and sticks used to form the drey were taken from the oak tree so that the squirrel never came down to the ground. Living twigs were bitten and pulled off sometimes with great difficulty. Most of the twigs selected were about three feet in length so that dragging them to the nest site, always in a forward movement, was often a cumbersome business, and quite as many sticks were dropped as were used. When a stick was dropped the squirrel would watch it fall but never attempt to retrieve it.

The drey was constructed by the squirrel trampling the twigs most actively, then using its forepaws to place them around itself, so that first a loosely woven ball shape was made; then the squirrel took twigs inside from the top and worked from the inside. Often it had quite a struggle entering and the squirrel had to push with repeated jerks of its head through the mass of sticks. After about two hours of collecting and placing sticks, the squirrel then came to ground to fetch great quantities of dried leaves and finally grass. These were deposited in the nest, the squirrel leaving immediately for another load. Intermittently more twigs were collected one by one and added, until the drey was ready for the squirrel's family.

The speed, dexterity and singleness of purpose with which the squirrel worked was equal to that of any bird.

during the following five minutes. The smelling forges a bond between cow and pup and during the three weeks' suckling period, the pup will announce its hunger by calling. Any cow in the neighbourhood may react to the calls but only its mother allows it to suck, after confirming its identity by smell. Hungry pups attempt to suck from any cow but they are brusquely rejected by any other than their mothers. The bond between cow and pup is thus one-sided and it is frequently the rule that a young mammal is undiscriminating while a mother will feed only her own offspring. An exception is the Mexican Guano bat. In the evening the mother bats leave their young clinging to the roof of the cave. As each female returns all the babies rush towards her and she feeds the first two to reach her. House mice sometimes have communal nests and share the suckling.

Parental care in mammals   While birds and mammals share similar problems of recognition of parents and offspring, and mammal species can, like birds, be described as altricial or precocial, mammals have evolved unique systems of parental care. The very word mammals is derived from the Latin *mamma,* meaning a breast, for the principal character of mammals is that they suckle their young with milk from the mother's mammary glands. One result of this specialized maternal feeding is a reduction in the role of the father. Whereas the majority of birds form a pair bond which lasts through the breeding season, with the male helping to rear the family, in the majority of mammals the female is left to rear the family by herself.

Antenatal and postnatal care   First, the female prepares herself for birth. Altricial mammals make a nest of soft vegetation and rabbits pluck hair from the belly to make a nest lining. The nest is usually set apart from any communal burrow or den. Precocial mammals retire to a quiet place to give birth. In both, before giving birth, the female licks herself thoroughly, particularly around the nipples and the genital region. After giving birth, licking is extended to the newborn infant. The birth membranes are cleared from its body, so ensuring the nose is free for breathing, and the skin is dried. Licking and the eating of the afterbirth are also important in keeping the nest clean and removing a source of attraction for predators. Even herbivores such as deer and cattle eat their afterbirths.

The female kangaroo and other pouch-bearing marsupials have extra antenatal chores. They build no nest because the babies are sheltered in a pouch. Shortly before birth the pouch is given a thorough licking. Only in 1959 was the manner in which the baby kangaroo finds its way into the pouch set beyond doubt. Just before giving birth the kangaroo adopts the birth position, sitting on the base of her tail, with her hindlegs extended forwards and tail passed between them. When the baby appears she licks it, then she leaves it to make its own way into the pouch. Although it is only three-quarters of an inch long, the baby kangaroo has strong front legs and takes about three minutes to make the passage. Once in the pouch the newborn kangaroo grasps a teat in its mouth. At first the teat acts as both anchor and source of food and the kangaroo does not emerge until it is eight months old. It is not weaned until another six months have elapsed.

No mammal other than the marsupials, or pouch-bearers, has such a difficult journey in search of its mother's teat. Among the true or placental mammals the newborn animal begins its search almost as soon as it is free of the placenta (afterbirth). Blind, almost helpless kittens, puppies and mice nuzzle their way along the mother's belly making side to side searching movements with their heads. During the search the mother may assist by moving into a convenient position or drawing them to her. Once at the nipple the babies stimulate the flow of milk by butting with their snouts or 'treading' with their forepaws.

During the time that young mammals spend in the nest, it is particularly important that they be kept clean. After clearing up the birth membranes and fluids, the mother continues to lick them at intervals, keeping them clean, establishing the bond with them and, by licking the anogenital region, stimulating them to pass urine and faeces which she ingests.

Newborn altricial mammals are also dependent on the mother for warmth but precocial young are born with a fully developed thermoregulatory system. Some follow their mothers immediately but others have a longer or shorter 'lying-out' period, during which they are left in a concealed spot while the mother feeds. Wildebeest, which live in very open country and are very mobile, rarely leave their calves, which can follow their mothers when ten minutes old and run as fast as they within twenty-four hours. Red deer calves, on the other hand, spend three or four weeks lying up apart from the herd, with their mothers staying nearby.

Carrying Altricial babies that are relatively helpless may be carried by their mothers. Some are carried as a

Caravanning of shrews

Certain European shrews indulge in what is called caravanning. The babies stay in the nest until seven days old, when they have the urge to wander. But before a baby does so it calls in a high-pitched squeak, sounding like a shrill whisper. Hearing this the mother runs to it and squats with her rump towards its face. The baby seizes the fur at the base of her tail in its teeth. Each of the other babies in the litter join in, holding the fur of the one in front. Then off they go, mother and five or six babies in line, all holding on so firmly that if the mother is picked up and held in the hand her family will dangle from her stern in mid-air.

There is nothing haphazard in this. All run in step, and should the caravan, for any reason, break up, mother shrew stops and by prodding and nuzzling, gets her family sorted out and off they go again. There is a two-way process because, in the event of a break-up, the babies wait with muzzles in the air, waiting to take hold once more.

The habit of caravanning was set beyond doubt when Dr. Hanna-Maria Zippelius, in Germany, photographed and filmed it in 1957-8. But it was already well-known in many parts of the world, notably south-east Asia. In the 1870's there were published accounts in *The Times of India* of musk-shrews caravanning. These musk-shrews often come into houses and when mother musk-shrew takes her family for an outing they usually run along the skirting, looking like a grotesque snake in the half-light.

matter of course, as with the baby koala on its mother's back, or the infant baboons, macaques and chimpanzees that start life hanging to the fur of their mothers' bellies. Later, these also ride astride their mothers' backs. Babies that normally stay in nests are often carried by the mother if they get separated from the nest for some reason or if disturbance leads the mother to move them to a new nest. Carnivores carry their young by the scruff of the neck whereas rodents are more likely to seize the nearest part of the body. When the young become displaced from the nest they utter distress calls, which may be ultrasonic (beyond the range of human hearing) in rodents and their mother comes to their rescue. Later, when they have learnt to walk their behaviour changes and a warning call from their mother sends them scampering from the nest to find individual hiding places.

Care in infancy The changeover from a diet of mother's milk to solid food is gradual. Young hoofed animals, such as horses, begin to nibble grass soon after birth and a long time before they are weaned. Grey seals are weaned, and abandoned by their mothers, three weeks after birth. They have to learn to feed themselves but can survive this difficult period by subsisting on the thick layer of blubber formed during the suckling period. Short suckling periods are the rule amongst seals and the only exception in the order Pinnipedia is the walrus. In this, suckling lasts for one year, probably because of the difficulty young walruses have in learning to find sufficient shellfish in the muddy seabed. For many flesh-eaters, where young animals have to learn to hunt, help is given by the parents providing solid food and hunting for their offspring during this critical period. A young Sea otter receives some solid food soon after birth but does not become independent for at least one year. In the later stages of childhood, it may die of starvation if a storm prevents its mother from getting sufficient food for both herself and her infant.

Carnivores, such as foxes, lions and domestic cats, bring food to their litters, either by carrying the whole carcase or by swallowing lumps of meat and regurgitating them in front of the youngsters. In some species, the cubs or kittens may be made to work for their meal by having to seize the meat from their parent and eventually live prey is brought to them. It is released and they have to chase and kill it themselves. In one instance, a tigress was seen pulling down a buffalo and leaving it for the cubs to kill. Being inefficient, they lost control of it and the tigress had to catch it several times until the cubs finally finished it off. In this way, the young carnivores learn what to catch and how to catch it. The parents do not actually teach them to hunt but present them with easily caught quarry on which to practise. Even when they have learnt to catch their own food young carnivores may stay with their parents for some time while their technique is perfected. They may hunt with their parents whose prowess will ensure that they do not go hungry.

Prolonged childhood is extended further in the primates, particularly in man, where learning of various skills and practice in social living are very important. Chimpanzees and gorillas learn what to eat and how to obtain it by watching adults; but their vegetable food is relatively easy to get and childhood is more important as a time for establishing relations with other members of the group. Baby monkeys spend the first part of their lives clinging to mother, or sometimes to an 'aunt', a particular female who is allowed to carry the infant. As it grows older, the baby is allowed to stray farther from the mother, returning for food and sleep or when frightened. It is tolerated by other adults and it can

Wild dogs' communal feeding

The Cape hunting dog is a powerful animal, up to 100 lb weight, its ears large and rounded, its short, sleek coat marked in blotches of brown, yellow and white. It hunts in packs of usually 12 to 20, preying on the large hoofed animals of the African savannah. Having made a kill, the pack consumes the carcase before trotting away.

These dogs have a highly developed social life in which the needs of the individual are subordinated to the needs of the group. This means sharing food, a well-fed member of the pack regurgitating part of the contents of its stomach for a hungry member to eat. The hungry member signals its needs by nosing the face and licking the lips of a satiated fellow-member. In this way, once the carcase meat has been wolfed, partly digested food is passed from one dog to another until all are fed.

There is no set breeding season and a bitch with pups is left behind in the lair when the pack goes out hunting; but she receives her share of the food when it returns. The pups begin taking solid food at about a month and they solicit the returning adults in the usual way. A pup is not necessarily fed by its parents, but by any adult with a full stomach.

Baboons are ground-living monkeys which for their own safety often have to run on all-fours. The female baboon instead of clasping her baby to her, as a tree-living monkey would do, at first carries it clinging to her belly. Later she carries it riding pick-a-back.

torment the senior males in a way that would draw a severe beating for an older monkey. The young monkeys play together, practising movement through the trees and engaging in mock fights. In Rhesus monkeys, at least, it seems that young males are more adventurous and go around in gangs while the females stay more with their mothers and show an interest in their baby brothers and sisters. The attitude of the mother is important to the development of her offspring. A good mother allows her baby to wander away to play but also cares for it. Bad mothers, however, either abandon their babies or are over-possessive. They may be overprotective and insist on clinging to the babies even when they are old enough to be playing with other youngsters. The result is that these monkeys grow up to be socially inadequate. They develop their own problems when it is time to mate and are, themselves, bad parents.

Although this last paragraph was written solely with other primates in mind, and is based on research into the behaviour of monkeys and apes, the similarities between what is said here and what obtains in the human family are too strong to be ignored.

Chapter Eight
Sleep and Comfort

Had Solomon lived today, instead of referring the sluggard to the ant he might have advised as follows: 'Go to the cat thou tensed-up victim, consider her ways and be wise.' For the cat, and probably all members of the Felidae, are masters of the art of relaxation; they are comfort-loving creatures, and indulge in sleep to a greater extent than most animals, with the exception of bats and dormice. A certain amount of comfort is desired by all creatures, human or otherwise; the most important ingredient is sleep, and there is a wide variety of other comfort behaviour.

Living implies an expenditure of energy, which is replaced from food eaten. Waste products result from this, some of which are eliminated by being passed to the exterior more or less directly. Other by-products accumulate in the muscles, causing fatigue and the need for rest and sleep.

Lions are noted for climbing abilities, but lionesses especially sometimes use the lower horizontal branches of a large tree to take a siesta.

Sleep and consciousness
True sleep occurs in fishes and is also found in the rest of the vertebrates, but the amount taken and the way it is manifested varies considerably. Sleep may be defined as an unconscious state or condition assumed at regular times in each period of twenty-four hours, these times varying with species. The amount of sleep taken also varies considerably.

170

Human beings spend about a third of their lives asleep. The domestic cat spends two-thirds of its time asleep and is doubtless matched in this by wild members of the cat family.

Although bats and dormice are traditionally accepted as symbolic of sleepiness, the sleep league for mammals is topped by the Two-toed sloth which sleeps for 20 hours in each 24, followed closely by armadillos and opossums, with 19 hours. Giraffes, elephants, horses and dolphins come at the bottom of the table with 4-5 hours sleep a day.

The conscious, or waking state can be defined as a state of alertness during which the animal is responding to environmental stimuli. Consciousness is associated with electrical activity in the **reticular system,** a diffuse network of nerve cells and fibres in the midbrain. During consciousness this system shows continuous electrical activity. In sleep this activity ceases, although the system keeps the animal alert and responds selectively to any incoming stimuli. These it monitors and either amplifies or reduces according to circumstances and, through the medium of the cerebral cortex, it either alerts the animal or lets it slumber. For these reasons it has been named the **reticular activating system,** or RAS. It might as well have been called the brain's watchdog. A hibernating hedgehog is in so deep a sleep that it has been described as almost the sleep of death. Yet its spines will be raised slightly every time a faint metallic click is made at distances of up to two feet.

Understanding of the workings of the RAS gives a clue to several forms of sleep, both usual and unusual, and at the same time reveals that there are intermediate stages between complete consciousness and total unconsciousness. One of these is **hypnosis.** The brain of a hypnotized man shows the pattern of electrical activity of full consciousness, although his behaviour is very different from that of normal consciousness. Animals can be hypnotized; the oldest known example is that of the domestic chicken, which can be hypnotized if it is held on the ground with its beak touching the ground. If one runs a finger from the tip of the beak through the dust in a direct line forward from the beak, the bird will remain immobile in this position without being held, as if it is fast asleep. After a while it rouses itself, shakes its feathers and resumes normal activity, as though nothing unusual had happened (see p 273).

An owl held in the arms, with its beak stroked gently from the base to the tip several times will, if then placed on the ground, appear to be dead. After a few minutes it will get on to its feet and fly away, quite unharmed. A weasel held in the hands and gently stroked from nose-tip to forehead goes limp and immobile for a few seconds, during which it can be swung from the fingers as if dead. These are only a few of many examples.

Forms of sleep

The amount of time animals spend daily in normal sleep varies enormously, and the degree of sleep equally so. It has long been said in Europe that the Common swift 'roosts in the heavens'. Recent research has shown that only when incubating does it sleep other than on the wing, which means that, except when accidentally grounded, a swift never touches down, other than in the breeding season. It is now thought that swifts, which alternate glides with short periods of wing-beats, take cat-naps while gliding.

The sleep of birds   It was not until 1963 that the first scientific observation was made on the sleep of birds, when Dr. Klaus Immelmann spent ten successive nights watching three ostriches in the Frankfurt Zoo. The birds rested between seven and eight hours each night, squatting on the ground with neck erect and eyes closed. The slightest noise, or a small beam of light directed on them, would cause them to open their eyes and be fully alert. Only for a few minutes each night would they lie on their side with legs stretched at right angles to the body and neck limp on the ground. At such times neither a bright light nor loud noises would rouse them. Moreover, only one bird at a time fell into this deep sleep. The rest periods varied from 5 to 102 minutes and were broken by the bird getting on to its feet to feed or empty the bowel, which it did on average seventeen times in a night. The deep sleep lasted on average only 9 minutes.

In general, this seems to be the pattern of sleep in birds, as shown by random observation of birds kept as pets in the house and of wild birds nesting on or near the house. A tame talking parrot in the house will utter words in a soft voice at almost any time of the night that one moves quietly about the house. Starlings nesting in a roof space can be heard moving their feet with a scratching sound, or making a low murmur, throughout the night by anyone in bed in a room under the roof. Moreover, it is not uncommon to hear a diurnal bird sing for short periods at various hours in the night. This is especially true where street lights are on all night, a notable example being the starlings that roost on buildings and twitter virtually throughout the night.

Animal Rip Van Winkles
Many of the small insect-eating bats of temperate regions spend about six months asleep, in hibernation. During the other six months they sleep all day and at dusk come out to hunt for insects for an hour. Then they retire to their roosts to sleep, or at least rest, until just before dawn, when they hunt insects for another hour. So these bats are on the wing for less than 600 hours in the year or less than 8% of their time.

The European Hazel dormouse (above) hibernates six months a year. During the remaining six months it sleeps by day and at night alternates activity with spells of sleep or resting in a drowsy state. Its active life occupies between 12 and 16% of its time.

The tuatara, the ancient lizard of New Zealand, hibernates lightly for half the year. For the rest it stays in its burrow by day, except to bask in the sun. Its metabolic rate is low, so it requires little energy to keep the vital body processes ticking over. It is so lethargic it even falls asleep while chewing its food. It is said to grow very slowly and does not breed until 20 years old; and it is estimated that it lives to

100-300 years. Some 90% of its time is spent sleeping or somnolent.

The Ground squirrels of Point Barrow, on the northern coast of Alaska, hibernate for nine months of the year. During the remaining three months they feed, breed, lay in fat and collect food for storing in their burrows. But they work at most seventeen hours a day, usually less. Their active life occupies about one sixth of the whole, or 16%.

You can't catch a goat asleep

A goat errs in the other direction from somnolence. It never closes its eyes. It is inactive for eight out of each twenty-four hours. If it lies down it may become drowsy but this is shown only by the head being slightly lowered and the ears drooping. A slight sound instantly alerts it.

Sleep patterns   Immelmann found that only one ostrich would fall into a deep sleep at a time, and this seems to be a constant pattern, not only for birds but for mammals also. Cows in pasture alternate grazing periods with resting periods throughout each twenty-four hours. During a resting period the whole herd will crouch on the ground with their heads erect, occasionally dropping their chins to the ground and closing their eyes for a cat-nap. From time to time one will roll over on to its side and lie with legs stretched at right angles to the body and neck limp, for up to five minutes, in a deep sleep. But only one will do this at a time. This same pattern has been observed in horses, as well as in herds of antelopes and zebra in Africa. It is always said that horses sleep standing, but from time to time one in a group will lie stretched out on the ground, and only when it wakes will another lie down, stretched out as if dead. Here again, the deep sleep lasts only a few minutes.

Sheep lie down to sleep but hold the head and neck erect and there is no sign that they lose consciousness. Lambs show true sleep, but only for brief spells of half an hour or so. Goats also merely lie on the ground, with head erect and eyes open. So far as is known they do not sleep, but merely fall into a state of somnolence, drowsing with a slight lowering of the head and drooping ears and eyelids. Any unusual sound immediately brings them back to the alert.

The Sooty tern, one of the graceful seabirds related to gulls, also seems to suffer from almost permanent insomnia. It breeds on islands, but for the rest of its time it is flying over the sea far from land. Unlike other terns and the gulls, its feet are not webbed, so it must pick its food from the surface of the sea without landing on it. Also, its plumage soon becomes waterlogged, so it either sleeps on the wing as swifts do or, like a goat, gets its rest in a state of somnolence. Even when on its breeding ground it has little time to make up for lost sleep.

Only super-predators, with no enemies to fear, can afford the luxury of extended periods of deep sleep. This is seen in household cats, although even these super-predators are not in a deep sleep all the time; they will immediately become alert if, for instance, a dog approaches.

Sleep mechanisms   Sleep has been extensively studied only within the last quarter of a century, and that mainly in human beings, by the use of the electroencephalogram. This consists of electrodes placed on the scalp which record the electrical activity of the brain on a paper chart. Using this apparatus, Nathaniel Kleitman, of the University of Chicago, discovered a few years ago that there are two kinds of sleep in humans. There is a light sleep during which the sleeper makes frequent movements of body and limbs. Then there is the deep sleep in which there is little movement except in the eye muscles and the extremities of the limbs. This deep sleep presents a paradox because the pattern recorded on the encephalogram is more nearly that shown in the wakeful state. Consequently, this deep sleep has been named **paradoxical sleep.** Investigation showed that we dream only during periods of paradoxical sleep, normally three or four bouts a night, which occupy 15% of our sleeping time.

Cats enjoy 15% paradoxical sleep except when they are kittens, when it is 90%; the amount drops to 15% by the time they are a month old. Birds show less paradoxical sleep than mammals, and other vertebrates, such as reptiles, show none at all.

During sleep there are physiological changes. The heart beat and breathing rate slow down and in most warm-blooded animals there is a slight drop in temperature. There is also a greater concentration of carbon dioxide in the

Wrens' communal roost
A man lived in a thatched cottage. There was a hole in the thatch where house sparrows had made a nest. One day in winter he was standing outside the cottage in the early morning and he happened to glance up. He saw a bird fly from the hole in the thatch. Then another and another. Transfixed he watched the stream of birds. There were 52 and the birds were Jenny wrens, one of the smallest of Europe's birds with a perky upturned tail and a total length of under 4 in, weighing $\frac{1}{3}$ oz. That was in 1956.

In the winter of 1962-3 wrens in similar number were found roosting in a nesting box. They were in three tiers all with their heads in the middle.

Wrens are mainly solitary birds, living well spaced out. In winter they often come together in clumps. The wrens of North America often build roost nests, which is unusual.

Even while the wrens are moving around, foraging during the day, they are prospecting for a suitable roost. It may be an old nest, nesting box, hollow tree or the like. While prospecting they keep in touch with their songs and calls which must, if only we could register and study them, add up to a language passed on from one to another.

We can only piece the story together from stray observations. It seems that one wren, of more dominant personality, eventually leads them to the chosen roost by calling or singing in a particular way. The signal is answered by other wrens farther afield. They are, in fact, whistling each other up.

As night approaches the leader is joined by one or more others. They hop about the vicinity of the

174

nest, among the ivy or wherever the roost may be. Others fly in. Then all fly to a nearby tree, always on the move, as though doing their best to draw attention to themselves as darkness begins to gather. Then the assembled company flies into the roost, one after the other in quick succession, drawn together from anything up to a mile around.

Over a century ago, the celebrated English naturalist John Gould noticed that hummingbirds sit on their perches at night with head drawn into the shoulders and beak resting along the breast. He found he was able to move them without waking them and that when he laid one on a table it remained in that position as if dead. Yet the temperature of the air around was 17-23°C (63-73°F), a comfortable room temperature for anybody.

Since Gould's day further research has been carried out which shows that the hummingbirds' nightly rest is more like hibernation than normal sleep. When an animal sleeps it uses less energy than when awake and active. When it goes torpid it saves even more energy. Hummingbirds, so restless and spritely during the day, save as much energy as possible during the night, except in the breeding season, when the female is incubating her eggs. A hen broods her eggs to prevent them from getting cold. A torpid hummingbird hen would be useless for this purpose, so she breaks her habit until her chicks hatch.

blood. The drop in temperature is very marked in some animals. Bats, for example, become virtually cold-blooded and must exercise their wings to regain the normal body temperature, or move about in the roost, before taking wing for a night's hunting. Hummingbirds also go into what is, in effect, a state of temporary hibernation, each night (except in the breeding season).

During the day bats **roost** in hollow trees, under bark, in caves or in roof-spaces and crevices in buildings provided by man. Without this habit of sleeping ensconced securely from predators, their chances of survival over the ages would probably have been slim, because they have a propensity for deep sleep. Other animals use less secure roosts, and it is noticeable that many of these, such as birds, have special roosting places where they gather as night approaches. There is sound sense in this when we recall how the RAS (reticular activating system) works by monitoring the sounds around and evaluating them. A simple example illustrates the value of this. We go to sleep undisturbed by the ticking of a bedside clock, because we are used to it. In other words, our RAS monitors this as a normal environmental sound and diminishes it. The sound of an alarm clock, by contrast, rouses us instantly, since the RAS has been stimulated by an unusual, loud and sudden sound, which it has then amplified.

Birds go to the same place to roost, night after night, where the environmental sounds do not vary significantly. Consequently they can sleep, lightly or deeply, and their RAS is not stimulated, except by unusual sounds, which may represent the approach of predators. We can compare this with the general inability of most people to sleep well the first night in a strange bedroom, as in a hotel. This is a safety mechanism and almost certainly would apply to a bird or other animal forced to spend a night other than in its habitual roost.

The function of sleep   Apart from the basic mechanism of sleep, which is now being slowly unravelled, the age-old mystery of sleep also involves its purpose or function. It has been called 'Tired Nature's sweet restorer'. Shakespeare elaborated on this idea:

Sleep that knits up the ravell'd sleave of care,
The death of each day's life, sore labour's bath,
Balm of hurt minds, great Nature's second course,
Chief nourisher in life's feast . . .

Sleep is a period of regeneration for mind and body, during which two of the chief physiological mechanisms are the production of a concentrated urine and an accelerated rate of cell division. The first means that body wastes are being more rapidly eliminated, clearing the toxins that cause muscle fatigue. The second brings growth and tissue renewal.

Very little attention had been paid to the subject of sleep in animals until about thirty years ago (except to take note of unusual positions adopted by some of them). Among observations made were those on seals. These usually sleep on land and soundly too. A seal pup asleep can be lifted without waking it and one scientist found he could lie on a sleeping Elephant seal without rousing it. In the water seals may sleep vertically with only the tip of the nose exposed at the surface. They can do this by inflating the throat like a balloon. This is called **bottling**. Seals can also sleep fully submerged, rising periodically to push the nose out at the surface to take breath.

Sleep in fishes   Before this time, also, nobody had any idea that fishes sleep. The discovery was made in the Aquarium of the London Zoo when somebody went in one night and switched on all the lights. He noticed then that the fishes in the various tanks were in unexpected positions and postures. Since then further random observations have reinforced the idea that fishes sleep. For example Grey mullet spend the daylight hours swimming in schools, evenly spaced and all with their heads in one direction. At night the schools break up and each fish goes to its own spot on the seafloor. The mullet are then well spaced out, with their heads pointing in different directions. Should anything disturb them they all swim up and form into the customary school.

Some fishes lie on their side to sleep, while others rest in the normal position. Some rest head-down in a vertical position. Flatfishes, such as plaice and flounder, lie on the seabed by day but float a few inches off the bottom when sleeping. Some young soles actually swim to the surface at nightfall and stay there with the body curved into a saucer-shape, with the fins round the margins of the body just breaking the surface.

There is another form of sleep which is known as hibernation, because it takes place in winter (*hiberna* is Latin for winter quarters). It is often called simply winter sleep. Like daily sleep it takes many forms, and of recent years zoologists have tended to differentiate between true **hibernation, winter rigidity** and **winter dormancy.**

Holding their breath
Elephant seals of the subantarctic seas spend much time on land, gathering in muddy depressions where a number of them lie on top of each other to sleep, forming a living pyramid. At times their wallows fill with water and then those underneath will be sleeping under water, so from time to time each must struggle out for air. Failure to do so will result in death by suffocation, and this does sometimes occur. Usually, however, the seal is saved by being able to go for as long as a quarter of an hour before coming out for air.

The Weddell seal lives round the shores of Antarctica, feeding on fish and squid. It spends most of its time in water and during the Antarctic winter when the seas are covered with a continuous sheet of ice it must sleep below the ice. It visits cracks in the ice to breathe, pushing its nose above the water, and if there are no cracks it will open breathing holes by sawing the ice with its teeth. At times the seal may have to travel for some time under water searching for a crack in the ice and then, like the Elephant seal at the base of a pyramid, it must hold its breath.

Until a few years ago it was a puzzle to know how it slept. Then it was found that it slept at a breathing hole, hanging more or less vertically in the water with the tip of the snout just above the water line.

When it is sleeping at a gap in the ice it makes use of what is called **bottling**. Other seals use the same method for sleeping at the surface, and so does the walrus. The seal blows its gullet out like a balloon, so it almost has a lifebelt round its neck. Exactly how seals do this is not known, but it is presumed that they swallow air for the purpose.

Fish sleep in pyjamas
Parrotfishes are brightly coloured and live among coral reefs on both sides of the Atlantic. Their teeth in both upper and lower jaws are fused, forming cutting edges that look something like a parrot's beak. They are among the few species of marine fishes whose sleeping habits have been closely studied. Many parrotfishes spend the day on the coral and migrate to underwater caves as night falls, the caves sometimes being a fair distance from their feeding grounds.

One parrotfish caused a mild flutter a few years ago when it was discovered that it did more than merely retire to a special bedroom. As night falls it begins to give off from its skin a slime or mucus that completely envelopes its body. At the front of this envelope is a hole guarded by a flap which allows water to enter and there is a hole at the back of the slimy envelope that allows water to escape, so the fish can continue to breathe while enclosed in what looks like a flimsy plastic covering. In the morning the parrotfish breaks out of its nightdress and resumes its normal activities.

Hibernation In true hibernation, which is restricted to warm-blooded animals, the internal preparations begin several weeks before the onset of cold weather. A store of fat is laid down and changes take place in the physics and chemistry of the body. One of the dangers of being inactive during cold weather is that tissues might freeze and ice crystals destroy the tissues. Many people, especially the aged poor, die of **hypothermia** in winter in temperate latitudes. It needs only a drop of a few degrees in the body temperature, with no auxiliary heat to counteract this, for the controlling centre of the brain to cease to be effective. In hibernants, provision is made for this in advance. There is a marked increase in the blood potassium. As the temperature of the air drops the body fluids become bound with chemicals of large molecular weight which lower their freezing point. These changes are all under the control of the hypothalamus, a small area of tissue on the underside of the brain.

Once the bodily preparations are complete the hibernant (often called the hibernator) awaits the critical fall in temperature. It enters a **hibernaculum,** often prepared in advance, and falls asleep, but instead of only a slight drop in body temperature, as in nightly sleep, there is a large fall. The normal temperature for the woodchuck or groundhog, of North America, is 34.9 - 40°C (94.8 - 104°F). During hibernation it falls to 4.4 - 13.9°C (40 - 57°F). At the same time the heartbeat falls drastically and the respiration rate drops to 14 from the normal 262 (cubic centimetres of oxygen per kilogram of body weight per hour).

For a long time it was thought that no bird hibernates. Then, in 1946, a poor-will was found apparently hibernating in the Chuckwalla Mountains of the Colorado Desert. Subsequent researches proved this to be true hibernation.

In cold-blooded animals the chemical changes in the body are not so closely linked to a narrow range of body temperature as in warm-blooded animals. At all times, their activity falls as the temperature falls and they become torpid. With the approach of winter in temperate latitudes they seek shelter, in the ground (snakes, lizards) or in the ground or in mud at the bottoms of ponds (frogs, toads, salamanders). Snails take shelter in the ground or under rocks; insects that over-winter as adults and spiders enter houses or other buildings, or shelter in dense foliage or under leaf litter. In these situations they are protected from the free circulation of air and its consequent chilling effects. They therefore merely become more torpid than usual, so saving energy. This is known as winter rigidity.

177

Some fishes appear to hibernate. The Basking shark of the North Atlantic loses its gill-rakers, without which it cannot feed, at the end of summer and retires to deep waters. The carp, a native of Asia introduced into Europe and the United States, moves into deeper water for the winter. There it forms groups of up to a hundred, in tight circles with all heads pointing inwards.

Winter rigidity   The use of the term winter rigidity is amply justified by the behaviour of the tench, a freshwater fish of Europe, which has also been introduced into the rivers of the United States. In winter the tench buries itself in the mud, and if dug out and thrown on to the bank it shows no movement, appearing stiff and dead, until it is tapped with a stick, when it promptly shows signs of life.

Winter dormancy   Finally, there is the third type of winter sleep often spoken of as hibernation, which occurs when the she-bear of temperate and Arctic latitudes dens up for the birth of her cubs. She is doing no more than sheltering from the cold, conserving her energy by becoming dormant with a lowered temperature and breathing rate. This is winter dormancy.

Summer sleep   Winter sleep, whatever form it takes, is an escape from the discomforts and dangers of lowered temperatures. It takes place when the ambient temperature falls below $10°C$ ($50°F$). In the tropics, especially in desert or arid areas, the reverse situation occurs. Animals are compelled to take measures to mitigate the effects of high temperatures, when the ambient temperature reaches above $25°C$ ($77°F$). Mammals sweat, pant or seek the shade; birds pant, but many others, especially small mammals and cold-blooded animals go into a summer sleep, or **aestivation.** The physiological processes involved, and the behaviour, parallel those involved in the various forms of hibernation.

One of the more specialized examples of aestivation, and the one usually chosen in the textbooks, is that of the African lungfish, which lives in stagnant or sluggish waters and breathes by gulping air at the surface. When the waters dry up in periods of extended drought, the lungfish secretes a leathery cocoon in a burrow in the mud at the river bottom and curls up in this to await the return of the rains. It can survive four years in this condition, during which it lives on its own muscle tissues and, as a result, becomes shrivelled in appearance, as if dead or mummified.

In all aestivation the animal remains inactive, so conserving energy, and subsists on food reserves stored in

Sleeping to escape the heat
In the opinion of some scientists true hibernation is found in those animals whose ancestors were originally tropical and migrated into temperate regions. This view is supported by the behaviour of the tenrecs, relatives of shrews and moles but having a spiny coat, that are found only in Madagascar, not far south of the Equator.

Like many other tropical animals the tenrecs fall asleep in the extreme heat of summer, a process known as **aestivation,** which is the opposite of hibernation. In the tenrecs' native country the temperature seldom falls below $60°F$, whereas most hibernating animals in the temperate regions go to sleep when the temperature drops below $50°F$. A tenrec taken to a Paris zoo showed how easy is the transition between aestivation and hibernation. It fell into a deep sleep on cool days in summer when the temperature dropped below $50°F$.

Aestivation is not so regular as hibernation and it is presumed that when animals migrated into temperate latitudes they slept in cold weather instead of hot and that in course of time this habit developed a seasonal rhythm in which the onset of hard weather is anticipated by internal changes.

Ladybird puzzle
Ladybirds are cold-blooded and they live on their own, except in winter, and then they group together under stones or logs, under loose bark, even under loose wallpaper in houses, where 50 or more may be found closely packed.

178

It used to be said that they 'clumped' in this way to conserve the heat in their bodies. There are two arguments against this. One is that ladybirds in hot countries will clump during periods of extreme heat; the other that ladybirds are also sometimes found in large aggregations on the surface of a post exposed to biting winds, frost and snow.

Dental grooming
Social grooming has long been known in monkeys and apes. In this, one individual will search through the hair of another using its fingers, picking out dried beads of sweat. These same animals are known to clean their own teeth, scraping the incisors and picking fragments from between their teeth using the thumb and forefinger as forceps. In recent years it has been discovered that one chimpanzee will clean the teeth of another in what is called dental grooming. Moreover, the chimpanzee doing the grooming may at times use what is in effect a tooth-pick, if necessary stripping the leaves from a twig, choosing one that has a pointed end that can be used for scraping or probing. If necessary, in the course of a dental grooming session, the chimpanzee using the tooth-pick may alternate between using the fingers only and using the pick. At such times as the pick is not in use it is either held in the lips like a cigarette or resting on the chest of the one that is being groomed.

the body. By its lowered breathing rate, and by remaining in a burrow or other shelter, it reduces considerably the loss of water from the body by evaporation.

Comfort movements
The general comfort of sleep, hibernation and aestivation, is reinforced by a whole gamut of daily behavioural tricks that can be grouped together as comfort movements. Two of the most widespread and familiar are those known as **grooming** and **preening**. The first is indulged in primarily by mammals, the second exclusively by birds.

Grooming   In Mediaeval English, a 'groom' meant a boy. Later, it meant a manservant, and in the seventeenth century was transferred to a man who curried and fed horses. In due course, the verb 'to groom' was applied to the activities carried out by mammals to keep the hair in good order and the body surface clear of parasites. The pattern of grooming is very varied. Dogs, especially the males, do no more than lick the forepaws and genitalia, nibble the fur with the

incisors and scratch with the claws of the hindfeet. At the other extreme, the domestic cat indulges in extensive 'washing', licking the fur. Rabbits groom their ears and in doing so take into the mouth an anti-rachitic substance consisting of oil from the skin irradiated by the sun. Some mammals have special claws or teeth that are used as combs for removing parasites and dead flakes of skin and to prevent the fur from becoming matted. Many mammals groom their young, and in some species, such as monkeys, mutual grooming or **allogrooming** has become a social function. Apart from grooming having a hygienic function it seems also to afford a pleasurable sensation and helps cement social bonds (see also Chapter 10).

The word 'grooming' has sometimes been applied to insects, although 'cleaning' is the more common usage. This occurs when a fly, for example, rubs its forelegs together and then passes them over its head or its wings, paying special attention to the antennae and the compound eyes.

Preening   'Preen' is an Old English word for a pin or brooch, but it soon came to mean to trim and adorn oneself, a most appropriate word to adopt for the way birds keep their feathers in good fettle, by cleaning, dressing and re-arranging them. If we watch one particular bird, unseen by it, throughout one day, it comes as a surprise how much preening it does, from a thorough-going toilet session to occasional applications of the bill lasting a second or two. No fashionable woman in a house full of mirrors could be more consistently fastidious.

In preening, a bird uses the cutting edges of the bill. The most frequent action, known as nibbling, involves taking the feather in the bill-tip and, starting at the base, working outwards towards the tip with tiny pecks. Or the bird may draw the feather through its bill in one continuous movement. Another trick is to stroke the feather with the closed bill, or quiver the bill against the surface of the feather. All are aimed at removing parasites or foreign bodies and stale preen-oil and re-arranging the feathers, re-setting any that are dishevelled.

In the course of preening saliva is worked into the feathers and also preen oil from a preen gland at the base of the tail, the only gland in a bird's skin. This keeps the feathers waterproofed and also makes them more effective as an insulating layer. Not all birds have a preen gland. Some, including herons, parrots, toucans and bowerbirds have patches of powder-down feathers on the body. These are

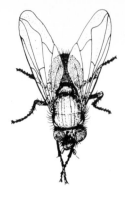

Even flies need to keep themselves clean, grooming their head with the forelegs and then rubbing the forelegs across each other to remove debris. The wings are cleaned in a similar manner with the hindlegs.

The feathers of certain patches on a heron's breast are readily friable, breaking up under the action of the bill to form a powder-down used in cleaning the rest of the plumage.

Cleaner shrimps

Skindivers, working at night in underwater caves, have seen sleeping fishes on the floor being cleaned, almost manicured, by shrimps. These same shrimps will also go over a lobster, cleaning its shell of small parasites. Even the voracious Moray eel, 4-5 ft or even 10 ft long, that will eat almost anything, will be rendered this service, several shrimps clambering over the huge body as the eel lies inert in a crevice. The eel makes no attempt to molest the shrimps.

There are cleaner fishes which pick parasites off the skin of larger fishes or clean up a wound. The cleaners have distinctive colours and live near some conspicuous object on the rocks, such as a brightly coloured anemone, which acts almost as a shop sign. Fishes in need of their attention present themselves at the shop and if two or more turn up at the same time they queue and take their turn. Large fish-eating fishes, including sharks, all of which open their mouths wide while one or a pair of cleaner fishes search their mouths and gill chambers for parasites, never eat their small benefactors, nor harm them in any way.

Cleaner shrimps and cleaner fishes perform the kind of service first noticed in the Tick birds of Africa, that clean the skin of wild cattle and antelopes.

Bears, in common with many other large mammals, rub themselves against solid objects such as tree trunks, posts or rocks. The purpose of this is not fully understood: it may serve to relieve a skin irritation but there are also indications that the animals derive pleasure or satisfaction from it.

friable and when pressed with the bill give off minute, dusty particles used in cleaning feathers. Powder down is especially important to herons for cleansing the fish slime from the feathers. These birds also have a comb on the third toe for scratching the head.

Scratching   In humans, as we all know, scratching may be a means to relieve an irritation of the skin, a displacement activity arising from a perplexity or from being thwarted, or merely something that produces a pleasurable sensation. The phenomenon occurs elsewhere mainly, if not entirely, among birds and mammals. It is most frequent in birds, but such study as has been given to it has been more concerned with the pattern of it than the function. It has been found that birds scratching the head, the only part they cannot reach with the beak, may do so directly or indirectly. In direct

181

scratching the foot is brought straight up, from under the wing or from in front of the wing. Indirect scratching of the head is achieved by bringing the foot from behind the wing, from over the wing or from between the wing and the body. Most birds, apparently, use the direct method, and this is regarded as the less primitive and more efficient of the two.

The pattern of scratching is said to be characteristic of a species and of its family, and some ornithologists purport to be able to establish the relationships of species by these patterns. They arrive at such satisfactory conclusions as this: that oystercatchers, avocets and stilts are more closely related to plovers than to other waders, because they scratch like them! Academically, this is all interesting, but for a bird the practical effect of scratching the head is that it relieves an irritation in a part of the body it cannot reach with the beak when preening.

Claws on the hindfoot (and hoofs in some hoofed animals) are used in a similar way, especially to reach the head, and sometimes to reach the flanks, as in the well known **scratching reflex** in dogs. If one touches a particular area on a dog's flank, the hindleg of that side is raised and agitated, the dog scratching the empty air. This is a pure reflex action, a response to the skin being touched and irritated by the slight pressure. Scratching with the hindfoot, on head and flanks, is a feature of the daily toilet of small mammals. Large mammals achieve the same end by using rubbing posts. These may be tree trunks, rocks or, in the tropics, termite mounds. The frequency with which these are used can be gauged by the paths or trails connecting them, which have been used by generation after generation and are well-known to beasts of prey as well as poachers.

Bathing, basking and wallowing Bathing in water is an activity associated more with birds than with any other group of animals. Some tropical snakes will lie immersed in water, possibly as a means of keeping cool, and some domestic dogs will lose no opportunity of splashing through water, but this seems not to be essential to health or well-being. In birds, **bathing** is a regular ritual, the pattern of which is innate. A fledgling will perform bathing movements on land when first confronted with a small bowl of water, and will often do so as it dips its bill in water to take its first drink.

Water birds bathe while floating on water, dipping the head and shoulders and throwing water over the back. This is followed by beating the wings vigorously on the surface,

Enjoying a cold bath
It seems almost to stand to reason that when a bird takes a bath in water it is doing so to clean its feathers. Yet the fact remains that birds bathe more, and more often, in cold and misty weather than at any other time. This suggests that the feel of cold water on their skins is a pleasurable sensation, and they certainly give every appearance of enjoying their bath.

In a small private zoo in England the birds in the aviaries were given fresh bath water every day. In summer it was often unnecessary to renew the water in the bowls for several days on end, showing clearly that the birds were using it for drinking and not for bathing. It was a different story in winter and on one particular day, when all around was frozen, a rook bathed with such enthusiasm that he emptied his bowl by the vigour with which he splashed in the water, in no time at all so that his bath had to be refilled. Altogether in one bathing session he used up several gallons of water.

On several occasions, the larger birds were seen to take a bath with enthusiasm and apparent enjoyment, yet the water froze to their feathers after they emerged from the bath and when they shook themselves their plumage rattled like a glass chandelier.

Yawning

We yawn when we are tired, during the day and especially in the evening, and we yawn especially on waking in the morning. A yawn makes us fill our lungs with air. This drives the blood from lungs to heart, so an increased supply of oxygen reaches the heart which stimulates it to pump the oxygenated blood to the muscles, so tending to reduce fatigue.

During sleep the rate of blood flow slows down, the blood itself becomes clogged with carbon dioxide and is sluggish. The morning yawn sets the blood coursing. Then we stretch, which squeezes the veins, sending more blood more quickly to the heart to be purified.

Fishes yawn mainly in the middle of the day, when they are active. If one of them has been still or moving very slowly for some time it will yawn as a prelude to moving quickly. This accelerates the flow of blood through its body, providing more energy.

A fish may yawn also when it is excited or when it sees an enemy or sees food, all occasions when rapid action is needed. The yawn tones up the muscles giving an instant supply of energy. A fish will also yawn when thwarted.

Another kind of yawn is seen when a hippopotamus opens its enormous mouth in a huge gape. This is a signal that it is ready to attack.

again splashing water over the back. Often one wing at a time will be more vigorously beaten. Land birds usually bathe standing in shallow water. The head and shoulders will be dipped, with the tail held out of water, then the tail and rear of the body are dipped with the head held up. After this the body feathers are ruffled and the wings flapped to produce a vigorous splashing. Intermediate patterns between these two are seen in semi-aquatic birds.

Although all these movements are rapid and vigorous, they are carried out in a way that obviates damage to the feathers. The aim seems to be to remove particles from the plumage and prepare the way for preening, and a noticeable next step is the drying of the plumage. Aquatic birds raise themselves from the water and flap the wings vigorously. Land birds leave the water and pass the long feathers individually through the bill, squeezing out the water. The body feathers are ruffled and the body shaken, throwing out droplets of water. The head and neck are rubbed against a solid object, such as the surface of a branch, to remove water. Cormorants return to land and hold the wings out, presumably to dry them (but see p 109).

Some birds **rain-bathe**; pigeons, for example, raise one wing vertically during rain, while tilting the body to one side. Others sleek the feathers as if to avoid too much wetting of the plumage.

Bathing is always followed by preening since a wetted plumage makes a bird temporarily vulnerable to predators, and no time must be lost before setting and re-arranging the feathers.

Before leaving the subject of bathing it should be recalled that this activity seems to be essential to the health of elephants, and the Water vole of Britain develops an eye-disease if denied frequent immersion in water.

Sometimes dust is used instead of water for cleansing. Dust consists of fine particles of dry soil or sand and **dust bathing** is a feature of the behaviour of a number of birds, while some large mammals do something very like it by rolling in dust. Birds that dust bathe usually do not bathe in water, although some species use both. Known dust bathers include larks, sparrows, sandgrouse, game birds, bustards, rollers, hoopoes, wrens, bee-eaters, hornbills and several others. Some owls and hawks will dust bathe, as will the rhea. The common pattern is for the bird to settle on the dry ground and, by using the bill, scratching with the feet, or rotating the body, create a saucer-shaped depression.

The feathers are then ruffled and dust thrown into them, usually by strong movements of the feet. The White-winged chough stands to dust bathe, applying beakfuls of dust deliberately to the feathers with movements reminiscent of preening. The rhea squats and picks up dust in the bill and throws it over the body. In all species, dust bathing ends in vigorous shaking of the body with feathers ruffled.

Some ornithologists see dust bathing as the equivalent of water bathing; others prefer to view the two activities as separate and dissimilar. Little is known of the significance of dust bathing except that it seems to produce satisfaction in the performer and may result in the removal of some surface parasites and possibly skin and feather debris.

Water and dust bathing are not found in reptiles and amphibians, or in any land animals other than birds and mammals. The reason is not far to seek: these others cast their skins periodically, a process known as **sloughing** in the vertebrates concerned and **ecdysis** in insects and other invertebrates. This achieves the end of cleaning the surface of the body.

Rolling in dust is a familiar activity of horses, asses and zebras, as well as domestic dogs. The action is always vigorous and appears to give satisfaction. It is followed by the performer getting on to its feet again and shaking its coat. The full significance of this behaviour is not known. Nor is the significance known of the rolling on carrion and dung that is such a reprehensible feature of the behaviour of domestic dogs. It can hardly be called toilet behaviour, although it may be a comfort movement insofar as it brings satisfaction to the performer.

The most universal form of behaviour to be included under the general head of comfort movements is that variously known as **sunbathing, basking** or **sunning**. Although this behaviour is widespread, little definitive study has been made of it, so it is not possible to discuss it except on the basis of random observations. We do know that there are few land animals, even those whose main activities take place at night, that do not sunbathe. Even some of the aquatic animals indulge in it. There are two species of fishes named after the habit, the Ocean sunfish and the Basking shark. The first of these, a giant marine fish with a disc-shaped body weighing up to half a ton, is sometimes seen at the surface, as if sunbathing. Recent observations indicate, however, that it does this only when moribund. The Basking shark also is seen at or near the surface. It is there primarily

Rabbits and rickets
Rabbits wash themselves paying especial attention to the ears. It was found on investigation, many years ago, that this was not merely a matter of cleanliness. When washing its ears a rabbit licks its forepaws. It runs the moistened paws over the surface of each ear, then licks its paws again and runs them again over the surface of the ears. So it conveys to its mouth small quantities of the natural oil from the surface of the ears. This oil contains an ergosterol which in the presence of sunlight forms vitamin D. Experiments showed that if a rabbit's ears were bathed with ether to remove the natural oil, the animal developed rickets. If the ears were left uncleaned and irradiated, even with artificial sunlight, the rickets could be cured. When, however, the ears were constantly washed with ether, to keep them free of the natural oil, no amount of sunshine, natural or artificial, would remedy the rickety condition.

Glorious mud

A dog taken into the woods for exercise may start to dig, which is a hunting or food-getting reaction, but as the smell of the freshly turned earth reaches his nostrils he is likely to turn on his back and luxuriantly roll on the fresh earth. He is more likely to do this if, in his digging, he turns over some rotting leaf-litter. He will also roll in the same way in bracken or other herbs, or on the putrid carcase of a rabbit or rat, or on a heap of dung. He will carry out the same actions in a puddle or a pond.

Whereas in human terminology he is befouling himself in the first instance and bathing (i.e. cleansing) himself in the last, both are sensual activities, which the dog finds enjoyable or pleasurable, and neither is functional. Both are accompanied by an appearance of ecstasy.

Horses and donkeys will roll on the back, with seeming enjoyment, and the dustier the ground the greater is the appearance of ecstasy. I do not remember having seen a cow or a bull behave in this way. The patterns of these indulgences vary with the species, but they also seem to be linked with the pattern of grooming. A dog seldom licks more than his paws and his private parts, but he grooms his head, back and flanks by rolling, or so we may presume.

to feed, although it may derive benefit from being exposed to the sun's rays. The sunfish of the fresh waters of North America is so named because it becomes more active in sunshine, more retiring when the sky is overcast.

Sunbathing is remarkably compulsive, which means no more than that there is an innate pattern of behaviour which is readily provoked into action by a strong stimulus. Take the case of a fledgling bird abandoned by its parents, which is taken into the house to be hand-reared. This young bird, relatively inexperienced in living, will show distinct liveliness when it first sees sunshine beyond the window, even although no rays penetrate the room. Given the opportunity, it will go as near the window as possible to gaze upon the distant, sun-illuminated landscape. It may even show signs of going into the sunbathing posture. The first time the bird is taken out into the direct rays of the sun it will sunbathe. That is, it takes up a position and a posture we associate with the word 'sunbathing'. It will crouch, breast to the warm earth, puff out the feathers of the body, droop the spread wings and fan the tail, exposing the maximum surface to the sun's rays.

Grown birds will do the same, frequently and persistently. They seem to have favourite spots to go for this, as if it were a ritual, although they will also readily sunbathe just where they happen to be. A bird will settle down in a crouching position, and a far-away look will come into its eyes, as if it were going into a trance. It fluffs its feathers, droops the wings, spreads the tail and seems to be oblivious to what is happening around it. One can approach a bird much more closely when it is sunbathing. If undisturbed, it may stay like this for an appreciable time, merely adjusting the position of a wing or the tail the more effectively to catch the sun. Often it opens its beak as if panting.

It has been shown that in some birds at least, the effect of sunbathing is anti-rachitic. That is, the oil from the preen gland, distributed over the feathers during preening, produces Vitamin D when irradiated by the sun, which is an antidote to rickets. On a subsequent preening, fragments of feather, bearing the vitamin, are taken into the mouth and swallowed.

Even typically night birds, such as owls and nightjars, can be compulsive sunbathers. They will sit or perch in the full sun, with the eye towards the sun closed, then deliberately turn around and 'do' the other side. On occasion an owl will sit with the head held so far back it looks as if decapitated, exposing its throat to the sun. In addition, an owl may fly to the ground and crouch with fully spread wings, almost flattened against the ground. In this position, it may turn its head through a hundred and eighty degrees to expose its face, with closed eyes, to the sun.

Mammals will lie out in the sun, showing every sign of enjoying it. Even a badger, one of the most nocturnal of animals, has its sunning places. Lizards will bask in the sun, flattening their bodies, spreading their ribs, and turning at right angles to the sun to expose the maximum surface to its rays. Snakes often sunbathe, especially prior to the breeding season, and it has been said that this is a necessity for the female snake, at least in the colder parts of the world, to enable her ovaries to ripen. Even frogs and toads, for whom a dry skin is the greatest hazard, will occasionally be seen taking advantage of the sun. Many insects seek the sun and seem to derive comfort and satisfaction from it; the most obvious addicts are grasshoppers and crickets. They flatten their bodies in a somewhat grotesque manner so that they look squashed and lifeless, yet they will spring into action immediately one tries to pick them up. Even lizards and

Elephant punkahs
The most obvious difference between the African and Indian elephants lies in the size of their ears, those of the African elephant being four times the size of those of the Indian. Whether this means that the African has better hearing has never been investigated. More important, the ears are a means of keeping cool: they radiate body heat.
Generally speaking tropical mammals have large ears, while those of related species living in polar regions are small. The rule is not invariable, the outstanding exception being the lion, which has small ears. But the lion is known to suffer from the heat and

Bathing and wallowing are essential parts of the routine of elephants. They splash and trumpet, and squirt water over themselves.

must use its behaviour to counteract it, for example, by lying in the shade.

An elephant's behavioural traits for keeping cool include wallowing and the use of shade, and of these two the second is the most likely to differ in the two species. Left to itself, the Indian elephant lives in the depths of the forest, and will remain there in the shade throughout the day. By contrast, the African elephant is typically an inhabitant of the savannah, with only limited shade.

It seems certain therefore that the difference in the size of the ears of the two elephants is directly linked with the differences in their habitats and behaviour. The African elephant needs the larger ears because it is more exposed to the sun.

grasshoppers have their chosen basking spots and will travel a long way from their customary stations to reach a rock or a post they have selected for the purpose.

In contrast to preening and the various forms of bathing, some mammals habitually spend much time **wallowing** in mud. A variety of explanations have been offered for this: that a mud-caked skin is the best protection from biting insects, that when the mud flakes off it carries away skin parasites, that it is a crude form of the mud packs used in beauty parlours. The pachyderms, like the thick-skinned rhinoceros, wallow in mud. So do pigs and deer; Red deer have traditional wallows in peaty ground. The real answer may be that they do it because they like it (that is, it gives them satisfaction, makes them comfortable and relaxed).

What are we to make of the male Père David's deer? After wallowing in mud it hooks up small pieces of turf on the tines of its antlers and skilfully throws these, with a toss of the head, on to its back, then finishes off by urinating on to its flanks and back in a quite incredible manner.

Chapter Nine
Movements and Migration

dispersal: plankton: marine dispersal: aerial dispersal: hitch-hiking (phoresis) — irruption: swarming — vertical migration — migration in mammals — migration of eel and salmon — movements and travelling: unusual journeys — bird migration: preparation: moult: *Zugunruhe*: migration routes: flyways: narrow front, broad front and loop migrations: formation flying: navigation: orientation: landmarks — homing: bees' orientation: salmon homing: homing in mammals

Greylag geese

Although we have seen that many animals have a 'home' — a territory, a home range, a nest or burrow which forms the focus of their daily activities — it is rare for an animal to spend its life in one place. At some time or another it travels, perhaps only a few inches, at one end of the scale, or perhaps thousands of miles annually at the other, as in the migration of some birds. The study of these movements seeks to answer two questions. Why do animals move from one place to another? And how do they know where to go? In other words, how do they navigate?

Taking the first question, an individual animal travels for two basic reasons: for feeding and for reproduction. But the individual animal's travel has wider implications for the species as a whole. The travelling of the individuals allows the species to spread through the whole of the habitat, thereby ensuring that the species uses the resources of its habitat to the best advantage; it also enables a species to invade new habitats. The invasions of recently-formed volcanic islands or sandbanks, and the recolonization of areas devastated by fire or other catastrophe are examples of successful spread into new habitats. However, as far as the

Formation-flying

In many parts of the world flocks of migrating birds herald the spring and the end of harsh winter weather. The most spectacular migrants are the large birds such as geese and cranes which spread southward in picturesque formations. Each bird flies slightly behind and to one side, so that the flock takes up a V or echelon formation.

Formation-flying is more than a matter of neatness or avoiding aerial collisions. Long-distance migration, particularly non-stop flying across oceans, is a strain on a bird's physical resources. Although powerful fliers able to cope with contrary winds, the endurance of large birds is limited by their ability to carry adequate amounts of fuel in the form of fat. For a small bird half its body weight of fat is only an ounce or two but for a goose or crane this would be several pounds — a severe handicap.

Therefore, the larger migrants use every method for conserving energy. A large bird wastes energy in the form of vortices or eddies of air trailing from the wing. The eddies are a natural consequence of an object being swept through the air at speed. In a V or echelon formation the birds make use of the eddies of the bird in front to give them a little extra lift. Over a long journey this can make a significant difference. The leading bird gets no such advantage, so it drops back after a while to let another bird do the hard work.

Grunion provide one of the world's fish spectacles. They live off the Pacific coast of North America and at certain high spring tides congregate in their thousands on the beaches to spawn. The young fishes hatch at the next spring tide and enter the water. When spawning the female digs her tail into the sand, a male wraps himself around her to shed his milt over the eggs as they are laid.

individual is concerned, travel is first necessary in its search for food. Young animals have to disperse from the nest or breeding ground to avoid competition for food with others that are being launched on a new life at the same time. This is as true for a mass of caterpillars on a cabbage as for Grey seal pups that leave the crowded rookeries on rocky islands around British coasts and spread across the North Sea, even reaching Icelandic and Norwegian coasts.

Dispersal

These dispersals by young animals are largely random movements. There is no definite route taken. The animals roam apparently without guidance save the necessity of finding a good source of food. For territorial animals, particularly hunting animals, there is also the necessity of finding a place where they will not compete with another of the same species. Young Tawny owls and weasels are chivvied from pillar to post by established elders and, unless they find an empty niche, their chances of survival are slight. In July, European moles are more often seen on the surface of the ground than at other times. They are young ones that have recently left the maternal nests and are searching for a place to settle but have been chivvied from the underground tunnel system by the adults. Indeed, it is a general rule that the dispersal period of young animals is a time of crisis in which many die. It is at least part of the reason that an excess of young are produced. The function of the excess of young animals is to make good the numbers of the population that have become depleted through deaths in the adult stock.

The extent of the wastage of young animals varies greatly. The single pup born annually to each Grey seal cow is sufficient to maintain the population. Small birds need to produce two clutches of half a dozen eggs to make good the results of predation and winter starvation, while among invertebrate animals production of young is enormous, to offset not only the numbers eaten by other animals but also the great losses from a wasteful system of dispersal. Many invertebrate animals have a stage in the life cycle that is particularly geared for dispersal. This is very often the larval stage, but not invariably so. In insects the larva, as a caterpillar or grub, specializes in feeding, while the winged adult is the one to travel long distances.

Marine dispersal The larva as an agent of dispersal is found particularly in the marine invertebrates. Sea anemones, starfish, bristleworms and oysters, to name but a

191

few, produce larvae quite unlike the adult form. The adults move slowly, if at all, but the larvae are minute, delicately shaped organisms that float freely in the sea. They cannot move far on their own accord but are carried vast distances in ocean currents where they make up part of the **plankton**, the collective name for passively floating animals and plants, derived from the Greek and meaning 'that which drifts or wanders'.

Barnacles are a prime example of a marine animal that is dispersed by a planktonic larval stage. The adult barnacle is firmly cemented to a rock, a pier, a ship's hull or even a whale, and is quite incapable of locomotion. A barnacle attached to a whale or to a ship may be an agent of passive dispersal. For example, the colonization of European waters by certain Acorn barnacles from the southern hemisphere was brought about by barnacles which were carried to Britain in World War II on ships' hulls. These barnacles spread through their new habitat by means of their planktonic larvae. The young larva, known as a nauplius, is released from its parent's body and floats in the sea. During this period it feeds on minute plants and drifts about until it changes into a cypris larva. The cypris is more mobile and it looks for a place to settle. When it touches a solid surface it explores it with its antennules, a pair of limbs on its head. It is looking for traces of a compound left on the surface by previous generations of closely-related Acorn barnacles and, if successful, it glues itself in place and metamorphoses into an adult barnacle. The abundance of barnacles on the shore and the severe problem of their fouling ships' hulls shows the efficiency of this dispersal mechanism, but it is achieved at the cost of countless millions of larvae which never find a home. They drift aimlessly through the sea until they die or are eaten.

Aerial dispersal   The hit-or-miss nature of planktonic dispersal is well shown by aerial plankton. As mentioned above, the dispersal stage of most insects is the adult. Many of the smaller species in particular are not strong fliers. They do little more than keep airborne and allow the wind to carry them about. Aerial plankton has demonstrated great efficiency in travel, however, as shown by the spiders and mites, as well as small insects, which have been collected in nets towed by aircraft at heights of up to sixteen thousand feet, and have appeared in unusual places many miles from home. However, as a means of dispersing a species, this method shows a certain amount of inefficiency. Charles

Many species of spiders indulge in aerial migration. The spider climbs to the top of a plant, plays out a thread of silk and waits for the wind to lift this to make it airborne.

192

Unusual insect swarms

Everyone is familiar with the migrations, often spectacular, of birds. Even fishes furnish examples of remarkable migrations, like salmon and eels. And some mammals, in the past at least, carried out migrations remarkable for the distances covered, such as the caribou, or for the numbers involved, as when the springboks in their millions trekked across the South African veldt.

Migrations are less common among invertebrates, although there is the striking north-south journey each year of the Monarch butterfly in North America. During the past thirty years evidence has been accumulating to show that quite a number of insects travel long distances, and in large numbers although we are still largely in ignorance of the reasons for these.

Even earlier there had been sighting of large swarms of insects, which alerted entomologists to a problem worthy of study. The Silver-Y moth, for example, not a large moth, crossed eastern England in such swarms in 1936 that the sound of their wings was audible as a distinct humming. No estimate was made of their numbers but a swarm of the same species in 1882 crossed the North Sea from Germany to England, a distance of 300 miles, and was described as resembling a dense snowstorm.

In 1862, a swarm of dragonflies observed over Germany was estimated to number two and a half billion; and when another crossed Belgium in 1900 the sky over Antwerp was said to have 'appeared black' with them.

Elton, the pioneer ecologist, has described how swarms of aphids and Hover flies were discovered by an expedition sledging on the Greenland icecap. There was nothing there for the insects to eat and, in any case, they were soon killed off by a blizzard.

There is, therefore, a huge wastage among dispersing animals. Aerial plankton forms the staple diet of swallows, swifts and martins and, at a lower level, of web-spinning spiders. But, for the continuance of the species, the wastage does not matter. For some species, the arrival of only one individual in a new habitat can lead to successful colonization, provided that individual is a female, pregnant or able to reproduce by parthenogenesis (see p 124). One such is the Bean aphid or Black fly. This pest arrives in the British Isles from the continent of Europe every summer and it has been calculated that if all the offspring of a single aphid survived they would equal, in one year, 'the weight of five hundred million stout men'. Winged aphids set about dispersal by deliberately flying upwards, attracted by the ultraviolet light in the sky. If they meet an air current, they are swept along in it; if not, they return to the ground.

Hitch-hiking   Air currents are not the only means by which insects and other small animals are carried about. Many hitch a lift on larger animals, a process called **phoresis** (from the Greek *phorein*, to carry). The Blister beetles are so called because their bodies contain cantharidin, a substance that causes blistering. They lay their eggs on the ground and the active larvae crawl up plants and wait for a solitary bee to arrive and collect nectar. While the bee is busy, the larvae grab some of the hairs on its body and are carried back to the bee's nest, where they creep into the cells and feed on the bee eggs and nectar. Those larvae that cling to a solitary bee have a very good chance of survival but most of the thousands in a brood are condemned because no bee visits their flower, or because they hitch a lift on the wrong kind of insect. A nasty example of phoresis is that practised by a fly *Dermatobia hominis* that lays its eggs on a particular species of mosquito *Janthinosoma lutzii*. The mosquito lands on naked human skin to suck blood and the heat of the skin straightway stimulates the fly eggs to hatch. The larvae drop off the mosquito, bore into the skin and feed on the underlying flesh.

The Blister beetle and *Dermatobia* are deliberate hitch-hikers and this mode of transport is essential for their development, but there are many occasions when other animals are carried by accident. This kind of transport has to be assumed when, for instance, new aquatic animals are found in ponds from which they were previously absent. For those that cannot fly or walk overland, it is usually suggested that they have been carried by larger animals. There is, however, little concrete evidence that this happens, except for rare accidental discoveries, such as Charles Elton's finding of an Orb mussel being carried on the foot of a frog.

Irruption Dispersal is particularly spectacular when it involves the arrival of a large number of individuals outside their normal range, as in the instance of the insects turning up in Greenland. This is known as an **irruption**. A number of animals are well known for irrupting. In general, such animals specialize in a particular food which for some reason becomes particularly abundant. The numbers of the animal rise spectacularly but, suddenly, the food supply dwindles. The animal faces a crisis which, for many other species, would result in massive mortality, but the irrupting species literally flee the area in search of new food supplies. They

Travelling ladybirds
The familiar ladybird, wholly beneficial to man, feeds enormously on nothing but aphids, and does so in the larval as well as the adult stage. Estimates vary as to how many aphids a ladybird will eat a day, but it is a considerable number. Aphids stay put most of the summer, so ladybirds have no need of speed to capture prey or escape from enemies. A few birds will occasionally eat them, but the red and black colours convey the customary warning. Since these beetles have little need of speed, one would not have thought they made long journeys. Yet in September thousands can be seen on the east coast of Britain, coming from the sea, landing on the sand dunes. Sometimes many fail to make landfall and their carcases are washed ashore by the thousand. There is another puzzle. On certain hilltops in California are well-known spots where ladybirds regularly hibernate (or go into winter rigidity!). The following spring they disperse and later in the year a fresh generation finds its way to the hilltops in the same spots. It is known that these insects are attracted to gather at a particular spot by the odour of those already there. Possibly this odour persists until the next winter, so drawing ladybirds to these same spots year after year. Whatever the explanation, it is a phenomenon useful to man. The Californian farmers gather the ladybirds and take them to fruit farms to feed on the insect pests.

The Snowy owl of the Arctic is very dependent on lemmings for subsistence, especially during the breeding season. Periodically there are population explosions of lemmings and Snowy owls close in to feast on them.

194

Migration roundabout

The Spiny lobsters of the Caribbean are nocturnal and solitary. At the end of summer they change these habits temporarily: they form giant queues and march steadily in parallel lines over the seabed in the sunlit waters of the Bahamas and the east coast of Florida. This extraordinary behaviour is presumably some form of migration. How it starts is only slightly less of a puzzle. As in all migrations, movements are preceded by internal physiological changes, and once one of the lobsters starts moving by day others see it and follow, each grasping the rear end of the body of the one in front.

William Herrnkind, of Florida State University, found that when he put non-migrating lobsters in a pool 6 feet across they just wandered about or formed queues in a desultory way. The moment he put four lobsters in the pool that had already queued, the rest joined them and marched steadily round and round the pool at the rate of six yards a minute, and continued so day and night for 33 days.

appear well beyond their normal range and some breed there, but generally they return home or die out. Not surprisingly, irruptions are most common among birds, but the most famous example of irruption is that of the Norwegian lemming, whose spectacular mass movements have given rise to absurd ideas about mass suicide and migrations to 'lost Atlantis'.

Irruptions of Norwegian lemmings are tied to a four-year cycle of abundance of their food. Similar cycles of abundance of several other small rodents living in the treeless tundra wastes of the Arctic lead also to irruptions of one of their predators, the Snowy owl. When the rodents are abundant, the owls lay larger than normal clutches and a higher percentage of chicks survive to become independent. Then, when the rodent numbers crash, the owls are faced with famine. Many fail to breed and others fly south to become an interesting, if temporary, addition to the bird fauna as far south as California and Bermuda.

From the point of human economy, the most important irrupting species is the Desert locust — the eighth plague of Biblical Egypt. Locusts are grasshoppers which have the capacity to breed explosively when conditions are more favourable, and their offspring fly in huge swarms in search of more food. The Desert locust affects a belt of land from Spain to Bangladesh, an area that is the home of 10% of the world's human population; in Ethiopia alone swarms once ate enough food in one year for a million people. The essence of locust control is to attack the breeding centre before the swarms form. This is particularly difficult in the case of the unpredictable Desert locust but other species, such as the Red locust, have been contained by international action because it is known that swarms arise in specific areas. Normally, the Red locust lives as an innocent solitary grasshopper over most of Africa, but in some areas of Central Africa seasonal flooding of grassy plains provides conditions for swarming. The local graziers burn off the old grass before the rains, so as floods subside after the rains, there is an expanse of bare mud that is just right for locust egg-laying. When the young hoppers (the wingless, immature stage) hatch out, the new green vegetation is springing up and they have plenty of food. A few months later, however, the grass is burnt up by the sun. The swarms of locusts have to move out of the plain, and they spread over Africa by the hundreds of millions. Since 1944, the Red locust has never been allowed to swarm and it is no longer a scourge.

Migration

Strictly speaking, a migrant animal is one that makes regular movements between two habitats, thereby exploiting two different resources. For people living in temperate countries, the best known migrants are the birds, such as swallows and warblers, that fly across the equator to winter feeding grounds. These birds are usually insect eaters. During the summer there are sufficient insects in temperate climates for them to raise their young, but they have to leave for warmer places before winter sets in and insects become scarce. Migration is practised by many other animals, although the movements may not be spectacular. Even a caterpillar descending a plant from the foliage where it has been feeding to the ground where it will pupate is performing a migration.

Vertical migrations   In any sizeable body of water there are regular up and down movements on the part of planktonic animals. These vertical migrations have been studied particularly in marine plankton. It is a widespread habit, yet scientists have not found a wholly satisfactory reason for it. Sir Alister Hardy has described vertical migration as 'the planktonic puzzle number one'. As a general rule, the animals rise towards the surface at dusk, then disperse somewhat during the night. At dawn, they gather again and actively swim downwards, reaching the lowest levels around midday. The extent of the migration varies from a few feet by protozoans, which can swim less than an inch per hour, to a few hundred yards by some crustaceans. Sir Alister Hardy considers that the migration is a means of travelling in search of food. The minute planktonic plants on which crustaceans feed tend to gather in patches at the surface. By migrating downwards even a few tens of feet the crustaceans meet currents which carry them several thousand feet horizontally, whereupon they swim back to the surface, chancing that they have moved into a good supply of food. Professor Wynne-Edwards takes a different view, suggesting that the animals gather at the surface to form swarms for mating purposes. Whatever the function, the nocturnal surfacing of plankton affects other animals. Shoals of herring which feed on the crustacean *Calanus* follow its vertical migrations, which is why fleets of drifters fish at night, and many seabirds feed at night when the plankton is near the surface.

Migration in mammals   Vertical migrations may take place on land as well as in the water. Mountain-dwelling birds, which migrate downhill for the winter, exemplify this.

Processionary caterpillars migrate nose-to-tail in single file. If the leading caterpillar is moved off-course by an interfering human, the caterpillars can be made to walk endlessly in a circle.

A kangaroo escapes from its enemies in leaps and bounds, in an action like that of a bouncing ball or a pogo-stick. Is this better than running away from its enemies as a quadruped would do? Two Harvard zoologists trained a pair of kangaroos to work on a treadmill, and they gave them face-masks which enabled the scientists to measure how much oxygen the kangaroo consumed.

Using calculations based on the speed and weight of a kangaroo and of its oxygen consumption, the two zoologists found that for speeds less than 11 m.p.h. the bounding and hopping is mechanically inefficient. The kangaroo then required more oxygen in relation to its speed than would a quadruped. At speeds of less than $2\frac{1}{2}$ m.p.h. a kangaroo hops most inefficiently, using its tail as a fifth limb. At these speeds the kangaroo used four times the amount of oxygen a quadruped of the same weight would use. Once a bounding kangaroo reaches more than 11 m.p.h. however, the oxygen consumption ratio falls. It seems that the faster the kangaroo moves the more efficient is its use of the oxygen it breathes. The kangaroo is able to use the oxygen direct for the production of energy, which is why it can achieve speeds of 24 m.p.h., or even 40 m.p.h. in short bursts.

However, land migrations are generally across country. Some mammals perform extensive migrations. The caribou of North America spend the summer on the barren grounds of the Arctic tundra. They migrate north in large herds in April and May and return in June and July to wooded regions in the south. In September there is a partial return to open country for the rut, after which they return to the woods for the winter. So regular is this migration that Indians used to intercept the caribou and drive them into corrals or kill them as they crossed rivers.

On the western coasts of North America, the caribou migration is roughly paralleled by the Pribilof fur seal. In November and December, the seals leave the waters around their Alaskan breeding grounds and swim south singly or in small groups. The males do not move far, but females and young may swim three thousand miles, as far south as San Francisco.

California is the scene of a migration that has become a tourist attraction. Gray whales spend the summer in the Arctic Pacific and travel down the North American coastline to give birth to their calves in the shallow coastal lagoons of California and Mexico. Before the ruthless hunting of the nineteenth century, Gray whales swam in herds of thousands. They were nearly wiped out, but under protection their numbers are recovering and, once again, they provide a magnificent sight for watchers on the shore. Long migrations are commonly undertaken by the huge baleen or whalebone whales of which the Gray whale is one. They give birth in warm water around the tropics, as the calves lack the insulating blubber layer of the adults. However, there is little food in these waters and the whales swim to cooler Arctic or Antarctic waters to feed on the masses of planktonic animals that flourish there in the summer. Their calves have by this time grown a blubber layer to protect them against the cold, and they can be weaned on the rich abundance of plankton.

Migration of eel and salmon The migration of fishes is particularly important to fishing industries; movements have been studied intensively by marking individuals with numbered tags. However, the migration of the European eel was worked out by painstaking detective work. Its breeding place had been a mystery since earliest times, until the Danish zoologist Johannes Schmidt began an investigation of the subject by catching eel larvae from all over the North Atlantic. The larva of an eel is called a leptocephalus because its discoverer, the German naturalist, Kaup, who caught one in the Straits of Messina, Italy, in 1856, not realizing what it was, thought he had found a new species of fish and called it *Leptocephalus brevirostris*. Schmidt found that the larvae became smaller as he progressed westwards from Europe, and he pinpointed the breeding ground in the Sargasso Sea in the western Atlantic. After hatching, the larvae swim to Europe, taking three years for the journey. On entering the rivers they change first into elvers (young eels) and later into adult eels. They feed in fresh water for several years then travel back downstream, sometimes wriggling overland from pond to stream, and finally swim back across the Atlantic to spawn and die.

The American eel is very similar to the European species and has a similar life history, although its maritime journey takes only one year. It breeds in the same area as the European eel, and there is a theory that both belong to the same species. In this theory it is suggested that larvae from American eels swim either to America or to Europe, but that those growing up in Europe never return to breed; they die on the way back. So the American population is continually replenishing the European population. This theory has suffered a knock from tests that show that blood serum from each population is chemically different and that there are really two species.

Fishes that, like the eels, spawn in the sea and swim up rivers to feed and mature are called **catadromous**. Fishes performing the reverse movement are called **anadromous**. They live in the sea and ascend rivers to breed. Salmon fall into this category. The Atlantic salmon of North America and European waters spawns in shallow freshwater streams and lays its eggs in depressions in the gravel bed which are called redds. The young salmon, when hatched, are called alevins. At first they feed on the contents of a yolk-sac; when this is absorbed they start to feed and grow into fingerlings. After two years, when they are four to eight inches long, the fingerlings become parr, distinguished by a row of dark thumblike marks on each side of the body. Eventually, after one to eight years, depending on water temperature and food supply, the parr develops into a silvery smolt which swims down to the sea. Here, it puts on weight very rapidly and after one to six years it returns to the same stream to spawn as a grilse, weighing up to sixty pounds. The life cycle of the six kinds of Pacific salmon is essentially similar, but whereas some Atlantic salmon survive spawning and return to the

Eels' magnetic compass
For over a century scientists have speculated on whether migrating animals are guided by the earth's magnetic field. As one ornithologist put it, years ago, when bird migration was being discussed, whenever a dead end is reached in any study of migration someone is bound to take the earth's magnetism theory from the shelf, dust it, and see if it works in this particular case (but see p 205).

Some fishes, it is known, are sensitive to weak electrical fields through their lateral line organs, the line that runs along the upper flank from gills to tailfin. Freshwater eels are sensitive to electric fields of about a millivolt per cm in fresh water. In salt water they would be sensitive to fields some four orders of magnitude weaker. But they respond only to magnetic fields perpendicular to their bodies. There is no response to fields parallel to their bodies.

A moving electrical conductor, such as sea water, flowing past the lines of force from the Earth's magnetic field, sets up an electric current in a direction mutually perpendicular to both the direction of flow and the magnetic field itself. If a fish is sensitive to this small current it should be able to orientate itself in respect to an ocean current.

This may be how eels migrating the long distance from the rivers of Europe and North America to the mid-Atlantic, and their larvae making the return journey, find their way.

Egg-laying is for female turtles a herculean task usually involving a long sea journey followed by a laborious crawl up a traditional sandy beach. They find their way to distant beaches across the ocean using, it is presumed, celestial navigation.

sea, as kelts or spent fish, coming back to spawn in successive seasons, all Pacific salmon spawn once only and die soon after.

Recent researches involving tagged salmon show that the European population feeds off Greenland; it is natural to ask why these salmon and European eels should make these lengthy, tiring migrations. The answer may lie in the migration of Green turtles. These spend much of their lives feeding on marine plants in shallow coastal waters, but they gather to breed at certain beaches. Turtles feeding off the coast of Brazil breed on the lonely island of Ascension some fourteen hundred miles away to the east. It is presumed that they navigate by means of the sun, but it seems unnecessary to travel so far to lay their eggs. A plausible explanation lies in the Theory of Continental Drift. According to this, at one time the Americas were joined to Europe and Africa, the bulge of Brazil fitting neatly into the Bight of Benin of West Africa. Over the course of ages, the continents have drifted apart with infinite slowness. So, once upon a time, Ascension was very close to the turtle feeding grounds off Brazil. As it moved away, the turtles continued to make the journey to and fro. Continental drift could also have resulted similarly in the separation of the breeding and feeding grounds of European eels and salmon.

Bird migration The migration of birds has been a familiar but mysterious phenomenon since man started consciously to take note of his surroundings. Birds were seen to disappear in autumn and reappear in spring or, in warmer places, to appear only during winter months. Clearly they were escaping the harsh winter weather that made their summer homes untenable. But where they went or came from was another matter. One suggestion was that swallows spend the winter sleeping at the bottoms of lakes. The mystery began to be dispelled when birds were marked with numbered rings or bands, and records kept of migrating birds seen at different places along the route. Bird banding, as it is called in North America, or bird ringing in Britain, and recording have become major hobbies undertaken by hundreds of amateur ornithologists, and the facts they have unearthed have answered many problems of bird migration.

In temperate latitudes, the most common migrants are the insect-eaters whose food disappears during the winter. The swifts, swallows, martins, warblers, hummingbirds and cuckoos fly south to the tropics. In the arctic or subarctic countries, the severer winters force birds south to the temperate regions. Geese and ducks, waders (shore-birds), thrushes and finches swell the winter populations of birds in temperate Europe and North America. This migration may not be complete. Some individuals stay at home all their lives, and survive the winters on their breeding grounds. Others move away in what is called a **partial migration**, presumably so relieving the pressure on the sparse winter food supply. Tropical birds also migrate to make best use of plant and insect food burgeoning after the rains.

That birds migrate so as to avoid a food shortage during part of the year is self-evident, but migration is not controlled simply by the difficulty of finding food nor by the onset of bad weather. The birds must move before food becomes scarce as they have to eat heartily and build a food reserve in the body before setting off. They must do this while food is still plentiful. The initiation of the migratory process, which includes the laying-down of food reserves, is controlled by the internal rhythm of the endocrine glands and reproductive organs. This rhythm is fixed genetically, although it may be modified to some extent by external factors such as food supply and weather. After the breeding season the reproductive organs of birds shrink and sex

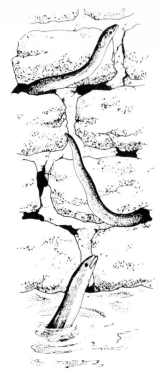

Eels climb brick walls
Sometimes in their migrations up rivers young eels meet with insurmountable obstructions such as dams and sluices, but they still seek to make their way upstream, by trying to climb the walls of the dam.
We had been told to look out for this around the brickwork of the sluices beside a mill. For a time nothing happened. Then an eel came up to the surface and swam backwards and forwards along the brickwork. Suddenly it made a swimming jump and wriggled its way up several bricks before falling back into the water. The next one to try was larger; it managed to push its head into a cavity between two bricks. Then, feeling with its tail, it found another hole just below where the head was lodged, and pushed that in. Disengaging the head, it stretched its body up and sought another hold. In this it was unsuccessful, and after a good deal of striving, partly supported by the

200

tail-hold, and partly held by pressure of the body against the bricks, it fell back into the water.

The best performance we saw was given by a medium-sized eel which not only put its head between two bricks in the vertical wall, but, after a struggle, disappeared entirely into the crack. Shortly it emerged, tail-first, let its tail dangle and, finding another hole with it, disappeared, this time tail-first. Later it re-emerged, only to drop back into the water again.

hormones disappear from the blood. Other hormones, particularly those from the pituitary gland at the base of the brain, are then secreted and these prepare the bird for migration. The rhythm of hormonal secretion is set by the change in daylength; captive birds can be brought into migratory condition by an appropriate regime of artificial daylight. Once in condition, the birds' departure is dictated by more immediate factors. A sudden dearth of food will drive them away, or they may linger if there is a spell of fine weather.

By the end of the breeding season, a bird's feathers have become badly worn, and they must be replaced by a new set so that the plumage is in peak condition for migration. However, this is not the rule with all birds. The European swift, for example, moults when it has reached winter quarters. Replacing feathers exerts a strain on the bird, so it must moult at the most convenient time.

Between breeding and migration a bird must allow time to lay down its food reserves. Fat is the preferred fuel for migration, as it is the most economical in terms of weight. One gram of fat yields over twice as much energy as one gram of carbohydrate. Some migrants almost double their weight before setting off, and such a fuel load gives even the smallest birds a huge range for non-stop flight. A warbler may carry only four or five grams of fat, but the distance it can fly is proportional not to the absolute amount of fuel carried but to the ratio of fuel to body weight, which in warblers is between fourteen and eighteen grams. Thus, we find warblers migrating non-stop across the Sahara desert, and the Ruby-throated hummingbird, comparable in size to the warblers, crossing the Gulf of Mexico. Hummingbirds' food reserves are laid down in less than one week and are consumed at a rate of about 0.5% of the body weight per hour. With a reserve of 50% of the body weight, these birds have sufficient fuel for their long non-stop flights together with a reserve in case of head winds or bad weather.

Before migration starts, many birds get an 'urge' to be on the move which manifests itself in a restlessness, often referred to by the German word *Zugunruhe* (journey unquiet). The gatherings of swallows before migration is a form of *Zugunruhe*, as is the fluttering against the cage bars by captive birds at the time of migration.

Migration routes Data from bird ringing give a good indication of the general routes that migrating birds take, and the locations of their winter quarters. European swallows spend the winter in Africa. The British population travels down through western France, eastern Spain and through Africa, to take up quarters in South Africa. They fly about 150 miles a day and take about one month. The Barn swallows of America, which belong to the same species, fly south via either Central America or the West Indies to South America, particularly Chile and Argentina. On their return, the swallows of both continents spread slowly northwards as the spring weather improves. Barn swallows reach Florida in early April and Alaska at the end of May.

A detailed analysis of ring recoveries shows that particular breeding populations of birds keep apart in winter quarters and avoid mixing. Both the Greenland and Svalbard populations of Barnacle geese winter in Scotland, but they stay a discrete hundred miles apart. There is often a leap-frogging effect in which more northerly breeding populations winter farthest south. There are six subspecies of Fox sparrow breeding down the west coast of North America, from Alaska to British Columbia. They migrate to the west coast of the United States, the Alaska subspecies reaching California while the Vancouver subspecies is sedentary. Other subspecies occupy summer and winter habitats in intermediate positions. The European Ringed plover shows a more marked leapfrogging. Arctic populations fly to South Africa, southern Swedish birds to West Africa, and Danish birds to southern Europe, while most British Ringed plovers stay put.

The precise form of migration is shown particularly well by visual and radar studies of moving migrants. Small birds migrate at low altitudes, usually under two hundred feet, and generally migrate by night, unseen except where they gather at lighthouses or flop exhausted on ships. Larger birds tend to fly higher, the record being some geese flying at twenty-nine thousand, five hundred feet over India, and they often fly during the day. The flocks of geese, cranes and other birds flying over in spring and autumn have played a part in the folklore of many countries. By recording the movements of ducks and geese through North America, F. C. Lincoln discovered the major flight paths, which he named **flyways.** These are also used by birds other than waterfowl.

Within a migration route, the birds may be guided or directed by **leading lines.** These are natural boundaries such as sea coasts, mountain passes and the edges of deserts. Two routes very important to European migrants are the Straits of Gibraltar and the Bosphorus, where the birds take

Young cuckoos fly blind
How on earth do young cuckoos find their way to Africa, from Europe, a month after their parents have left? The question must have been asked thousands of times in this form, without receiving an answer.

The parent cuckoos reach Europe in spring. The female lays her eggs in the nests of other birds, leaving the foster parents to rear the baby cuckoos. So the youngsters never see their parents, who migrate south in July while their offspring are still being fed by foster parents. In August the young cuckoos fly south.

What we do know is that adult cuckoos kept in cages and aviaries begin to get restless at about the time that free cuckoos are winging their way south. What is more, cuckoos so confined beat themselves on the south side of their prison, with flapping wings, showing every indication of an overwhelming impulse to fly south. Moreover, baby cuckoos hand-fed by misguided but generous people, who think the young birds have been abandoned and forsaken, do exactly the same approximately a month later. If anything, young cuckoos make even more frantic efforts to go south, so that their benefactors are compelled, out of common humanity, to release them before they do themselves major injury.

The young cuckoo, when released, may not immediately head south. In the confusion of being given its freedom it may even fly north a short way. But it soon gets its bearings and heads south again.

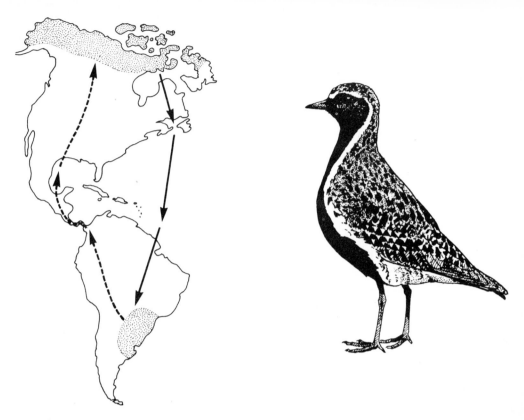

Map showing migration routes of American golden plover: the birds leave their breeding grounds in the Arctic and fly down over the Atlantic to their wintering grounds in South America; the return journey the following spring is made almost entirely overland.

advantage of short sea passages to avoid a lengthy flight across the Mediterranean. The coast of Holland, Belgium and northern France is a flyway for many small birds. Scandinavian chaffinches may cross to Britain across the North Sea, like the Vikings, or fly down the coast of Europe and cross at the narrowest point, like the Vikings' predecessors, the Angles and Saxons. Streaming along a flyway is known as **narrow-front migration,** as opposed to **broad-front migration** in which the birds straggle over a vast territory.

A few birds return on a different route from that of their outward migration. American Golden plovers fly from the tundra of Alaska and northern Canada across to Labrador. Then they fly over the sea for twenty-four hundred miles to Brazil, and through South America to the Argentine at speeds around 50 m.p.h. They return up the west side of the continent, via Central America and the Mississippi Valley. Strangely, young Golden plovers leave the tundra after their parents and fly south on the inland route taken by the north-bound adults. **Loop migrations,** as these are called, are more often used by seabirds taking advantage of prevailing winds. Giant petrels nesting on islands of the Southern Ocean (the seas bordering the Antarctic continent) sail around the world on the almost continuous westerly winds that prevail there, and the Short-tailed shearwater performs a figure-of-eight flight around the Pacific Ocean. It leaves its colonies around the Tasman Sea and flies out to the north of New Zealand, then turns up to pass close to Japan. The course keeps near the land around the top of the Pacific basin and down the west coast of North America before turning across the tropical Pacific. Off the north-east coast of Australia, the incoming track intersects the outgoing path in the East Indies, and the shearwaters fly down the Australian coast to their colonies. Throughout this immense twenty-thousand mile journey which fills in the time between breeding seasons, the shearwater is being assisted either by tail winds or by winds coming from the equator.

Navigation On any orderly migration, as opposed to a mere spreading or dispersal, an animal must hold to its course. A sense of direction or faculty for **orientation** is possessed by many animals. A limpet on the seashore will, as we have seen, make its way back to its own rock, where its shell has worn a groove to make a watertight joint as the tide falls, and an ant makes its way unerringly back to its nest with food for its companions. Some ants find their way by means of a chemical trail but others, such as the Wood ant of Europe, and many other animals, including honeybees, navigate by means of the sun's position. The animal steers by keeping the angle between the sun and its course constant. If, for instance, an ant is picked up, carried several feet and released, it continues on the same course and consequently misses its nest. If it is kept in a dark box for a few hours, it will when freed set off keeping a course at the original angle to the sun. But the sun will have moved through the sky so the ant will walk in the wrong direction. A simple 'sun-compass' is, therefore, only of limited use. It is adequate for short journeys but has limitations for long distance migration. Migrant birds are sometimes blown hundreds of miles off course, but when the weather improves they set off in the direction of their destination rather than continuing blindly on a set course as would an ant. The advantage birds have over ants is the possession of an extremely acute sense of time. This 'internal clock' allows them to relate the sun's position to the time of day and to alter the angle between the sun and their course as the sun moves through the sky.

How birds find their way over the trackless distances of the sea, often with surprising accuracy, is a problem that has

Unusual journeys
Although the general rule is that animals are pinned down to their territories, there are odd individuals that make unusual journeys. Several birds and insects have crossed the Atlantic from North America to the British Isles, the insects including the famous Monarch butterfly, but there is always the suspicion that these may either have done part of the journey resting on a ship or may have island-hopped using Greenland and Iceland.

Perhaps the most extraordinary wanderer was the hippopotamus known to people in South Africa as Huberta (but who proved to be a male, as was discovered after 'her' death). This hippopotamus wandered for no known reason a distance of a thousand miles from East Africa to South Africa. His progress was documented all the way by reports in local newspapers.

Another wanderer was the leopard found enbalmed in ice on the top of Mount Kilimanjaro, and on the continent of Antarctica seals have been found dead 100 miles or more inland, having wandered well away from their normal habitat.

There is a bat on Hawaii that so closely resembles the Hoary bat of North America which is a habitual migrant, that some years ago D.H. Johnson, of the Smithsonian Institution, suggested that it represents the descendants of a Hoary bat that made the ocean crossing some thousands of years ago. The distance involved is far in excess of the hundreds of miles that Noctule bats in Europe are known to migrate, but remembering Huberta almost anything becomes possible in the way of animal wanderers.

Homing dogs

Dogs are often reported as returning unaided to their old homes many miles distant. A typical instance concerns a foxhound, sent as a gift from one Hunt Pack to another 110 miles away. The hound was taken with others in a closed van but never saw the inside of the new kennels because as he was being off-loaded he bolted. He turned up again in his old kennels a month later.

The usual explanation is that tens of thousands of dogs each year are moved from one home to another. Most settle down happily in their new abode. The few that do attempt the journey back wander at random. Some get lost, a very few happen by chance to get back to their old home. The hound in question took a month to travel 110 miles, or three miles a day average, when it could cover 20 to 30 miles a day when out hunting. On the other hand, it had to cross two broad rivers. It also had to feed itself, which would account for further delays. All this story proves is the hound's persistence.

A more convincing story comes from Virginia, about a trained pointer taken in the car trunk at night from its old home to the new, a distance of 30 miles. After several days it was turned loose at noon, and half-an-hour later it disappeared. Three days later its former owner appeared with the dog, which had taken less than 16 hours to return home. Moreover, enquiries made of people living in the district showed that the dog had not gone home by the road along which it was brought but followed a parallel track.

still to be completely unravelled, but from experiments on captive birds and observations on wild birds it seems increasingly likely that birds navigate mainly by the sun or stars.

There is plenty of evidence that birds use celestial clues for navigation. If migrant birds are kept captive they flutter against the bars of the cage or perch facing the direction in which they would fly if free to do so. They can be fooled into orientating in the wrong direction if the sun's position is apparently altered by using mirrors, and if placed in a planetarium, a bird will take up position according to the cues given by the 'stars' on the domed ceiling. As a 'natural experiment', Adélie penguins were taken from their rookeries and released on a flat snow field. After wandering around for a while, presumably while they got their bearing, the penguins set off unerringly for home. But when clouds covered the sun they became lost and waddled about at random. The accuracy of celestial navigation was demonstrated by a Swainson's thrush which flew between two points in eight hours. A straight line between the two points measured four hundred and fifty miles, and the thrush's actual flown track measured four hundred and fifty-three miles, so its steering must have been extremely accurate.

Birds cannot depend on having clear skies while on migration, yet they do not appear to be too much deterred by overcast conditions. They are unwilling to take off in dull weather, but if overtaken by clouds they will keep going on the right course. Such observations suggest that there is more to bird navigation than orientation by celestial clues; they must have other methods of steering a course.

In one remarkable experiment homing pigeons were able to find their way to within two hundred yards of their loft while wearing 'frosted' contact lenses. Clearly they could not have been steering by the sun. The probability is that they were steering by the Earth's magnetic field, using some kind of internal magnetic compass. The supporting evidence for this is that homing pigeons carrying small magnets are completely disorientated if the sky is overcast. On a cloudless day they can steer by the sun and the magnets do not worry them. As well as pigeons, European robins and Ring-billed gulls have been shown to be influenced by magnetic clues, and the phenomenon is probably widespread.

There is still one vital piece of information missing from our knowledge of bird navigation. We know that a bird can steer a straight course by the sun or a magnetic 'compass', but how does it know which course to steer? For instance, a Manx shearwater was taken from its burrow on Skokholm Island, Wales, and carried to Boston, Massachusetts. It was back in its burrow in twelve and a half days, having flown three thousand miles across parts of the Atlantic not frequented by this species. Similarly, a Laysan albatross flew from the Philippines to its nest on Midway Island, a distance of forty-one hundred and twenty miles, in thirty-two days. There must be some sort of 'map' or other source of information that tells a bird in which direction lies 'home'. It has been suggested that it uses the height of the sun at noon to calculate latitude and longitude, rather in the manner of a human navigator using a sextant, but experiments designed to prove this have failed. Whatever form information for navigation takes, it must be carried in such a form that it can be passed from generation to generation. How else could young American Golden plovers or European cuckoos find their way to their winter homes after their parents have already departed?

Homing   Celestial navigation is probably used also by migrating whales, turtles, salmon and others as well as by birds. A feature of many migrating animals is that they return to exactly the same place year after year, a phenomenon known by its German name of *Ortsruhe*. This is particularly noticeable in colonies of burrow-dwelling seabirds such as shearwaters and puffins. Banding shows they have a remarkable fidelity to a particular burrow. If an individual changes its burrow it is probably because it has lost its mate and has been attracted to another in a nearby burrow. The accuracy of homing to a burrow is shown by Wilson's petrels, which nest in the Antarctic and can dig through newly fallen snow to reach their burrows. Homing in to such a specific place is probably not possible by means of celestial navigation; other cues are needed. Like the ship's navigator who needs a pilot to bring the ship into dock, a bird needs very specific information for homing. From what is known of homing or racing pigeons, landmarks around the home loft provide the basis for final recognition and guidance. Homing pigeons have been used for carrying messages since the days of the Ancient Egyptians and Greeks and frigatebirds are similarly used by Pacific Islanders. The practice is based on the bird's ability to return home after being released from a distant point. There is a great range between individuals' powers of homing, and pigeon racing is

Dying to reach home
How often does a bee fly in at an open window to inspect a bowl of flowers on a table in the corner of a room! Having gathered any nectar or pollen available the insect heads for home, making a bee-line for the hive, which more often than not takes it to a closed pane of the window. There it buzzes up and down the glass, this way and that, until someone takes a postcard and gently edges it to the open window and liberty.

We may say the bee hasn't the sense to go out by the way it came in. It is not a matter of 'sense'. The bee navigates by the sun. It can take its bearings by the sky even if there is no more than a small blue patch in an otherwise total overcast. So taking its bearings indoors presents no problems.

Should nobody be present to play the Good Samaritan the bee will succumb to its frustrated flutterings in little more than an hour. Its pathetic corpse will be found later on the interior window sill. This is not a matter of exhaustion. A bee can fly for miles without taking food and still have reserves of energy. It will have died form **auto-toxicity** (self-poisoning) due to stress.

Some years ago, in a series of refined experiments, some American scientists established that even a bee can succumb to stress much more rapidly than to exhaustion or starvation. In the circumstances described here, the fluttering bee on the closed window pane has only one endeavour, to go home. Such is its instinctive method of navigation it must do this by the shortest route. When it cannot do this its whole nervous system is thrown into confusion, with speedily fatal results.

Incidentally, this direct navigation, guided by the sun or the polarized light in a patch of blue sky, was unravelled and its mechanics made known a mere thirty years ago. The common man paid tribute to it, recognizing it from rule-of-thumb observation, and has been speaking of 'making a bee-line' for centuries.

Hunting wasps dig shafts in sandy soil which they provision with caterpillars for their grubs. They find their way back to the holes by memorizing landmarks, such as pebbles and pieces of twig. If these are re-arranged (below right), a returning wasp is momentarily lost and must reconnoitre to re-find the hole into its shaft.

based on selecting the small proportion that will home at speed over long distances. The pigeons home only to the place where they were reared, and they home best after they have bred there. To some extent their performance improves with practice, and it seems that conspicuous landmarks aid homing.

It is difficult to experiment with landscape cues in the homing of birds because the landmarks are so large, but insects lend themselves admirably to such tests. There are many solitary bees and wasps which construct a small cell or burrow which they provision with nectar or insects for the growing larvae to feed on. Provisioning requires several foraging flights in search of food, and the insect must be able to return unerringly to its nest. As it emerges it flies to and fro in a wavering flight for a few seconds, then sets off on a straight course. The wavering flight carries the bee or wasp over an area of a few square yards, and while on this flight it memorizes landmarks. This was first demonstrated by Niko Tinbergen with the Digger wasp which burrows in sandy soil. While the wasp was in its burrow, Tinbergen placed a ring of fir cones around the entrance. On emerging,

the wasp performed her wavering flight and went off to forage. Meanwhile, the ring of cones was shifted a few inches from the nest entrance. When the wasp returned she homed into the centre of the cone ring and was quite unable to find her nest.

Different cues are used by homing salmon, which return to their natal stream to spawn. Tagging of salmon smolts on their journey to the sea shows that when they return they come upstream and pick the exact branch stream from which they originated. At each fork they make the correct turn, but if their nostrils are plugged they are lost. The salmon are smelling their way upstream, although what it is they smell is not known. It may be the smell of young fry and parr that are their younger siblings and cousins, or it may be the smell (taste) of the water. There are professional water tasters working for water companies who have an almost uncanny ability to say, by tasting it, not only which river but which part of the river a sample of water came from. It has to be taken from a river with which they are acquainted, but the way the taster can, by taking a sip, say which of a dozen rivers the water came from, shows how refined this particular sense can be. All that is then needed for the salmon is a good memory of the smell (or taste?) of the water in which it was hatched.

A shark accompanied by its bevy of pilotfishes. These small fishes were named for a fanciful idea that they guided sharks on their wanderings. The fact is they use the shark's slipstream to facilitate their own swimming.

Right: The African mouth-brooder is a conscientious mother. After hatching the baby fishes stay in her mouth for some time before venturing out. Even then they stay near her and, if danger threatens, she signals to them with fin flicks and colour changes, at which they dash back to her mouth. She picks them up or allows them to swim in, in response to their touch.

Below: Unlike most mice and rats, baby Spiny mice are precocious. They are born with eyes open and a good coat of fur, although without the prickly hairs that give them their common name. This baby has managed to creep from the nest when only a day or two old and its mother is taking it back. Rodents commonly retrieve their young by carrying them in their mouth.

Left: Birds' nests can be very intricate. The male Redheaded weaver uses pliant green twigs to weave a nest; other weavers use grasses. When the ring of vegetable matter has been completed the male weaver displays to attract a mate before completing the nest, by weaving a tightly woven ball perhaps with a porch hanging from the entrance. Fixed to the tip of a twig, the nest offers protection to the eggs and chicks from enemies other than tree snakes. Its thick structure insulates them from the sun's heat and the coolness of tropical nights.

Below: Care of the offspring reaches a peak in the social insects. Ants, bees and wasps feed and tend their larvae and these African Ponerine ants are laying out their pupae in the sun after the nest has been flooded by rain. The ants also dug out the sodden earth from their underground nest and rebuilt it.

Right: Wolf spiders carry their families on their bodies. The spiderlings are free to roam over their mother's body but are kept clear of her all-important eyes. Part of the silken egg cocoon can be seen still attached to the spinnerets. If the spider is deprived of her cocoon she will search frantically for it. Apart from the protection received from being carried by such a fierce mother, the eggs and spiderlings benefit from being exposed to the sun's warmth and being taken into shelter during inclement weather.

Below: Sunbathing is enjoyed as much by animals as by human beings. This Collared dove seems to be oblivious of all else as it basks in the typical high-intensity posture with one wing raised. It adopts a similar posture in rain bathing. The bird gives every impression that it is enjoying itself but the function of the sunbathing is not clear. It may be that sunlight turns oils on the feathers into Vitamin D which the bird ingests during preening.

Zebra showing two kinds of sleep. Below: A line of zebra dozing lightly. The eyes are half closed and the ears are in a relaxed position but the animals are still alert. Short periods of rest occur throughout the day. Left: Occasionally one of the herd slips into deep sleep stretched out on its side. The single kongoni in the background is alert and its reactions to approaching danger will wake the zebra. Members of the horse family spend more time dozing than do ruminants such as cattle or kongonis.

For hundreds of years the disappearance of swallows in autumn and their reappearance in spring was a mystery. Fantastic theories were propounded to explain their absence. Knowledge that European swallows fly to Africa and back to avoid the rigours of a northern winter is no less incredible than the early theories. These swallows are resting against a backdrop of flamingos after a journey which is remarkable for its distance and the precision of navigation. When the swallows return to Europe each will make its way to the site of the previous year's nest.

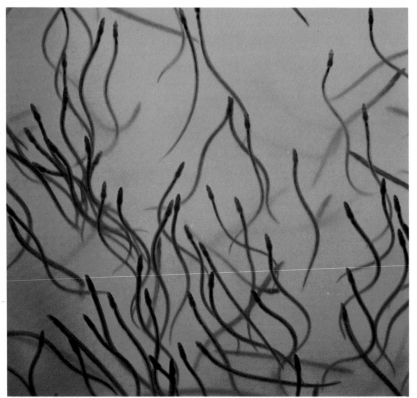

Left: Both American and European eels breed in deep waters under the Sargasso Sea. The baby eels migrate across the Atlantic to the rivers where their parents came from. As elvers, shown here, the young eels swarm in vast numbers up-river at one particular season.

Below: The salmon's migration is the reverse of the eel's. The adults spawn in shallow river pools and streams and young salmon migrate to the sea after several years in fresh water. They return as magnificent adults with their bodies in peak condition. They reach the mouths of their natal rivers probably by using celestial navigation, but find their way back to the actual pool where they hatched by smell. Their sensitive nostrils pick up minute quantities of scent which probably comes from the bodies of the next generation of young salmon living on the spawning ground.

Below: Although sometimes called White ants, termites are not related to true ants but they have a similar way of life. An important difference is that after mating, the male termite does not die but remains as a 'king' in the new nest. The picture shows winged adults of both sexes setting out on the mating flight. The termites' nest is a huge mound riddled with tunnels where the royal pair or pairs live with countless workers, soldiers and developing adults. The workers gather plant food which is digested with the aid of Protozoa in the gut or through the use of specially cultivated fungi and bacteria.

Overleaf: An example of mutual aid: three Sea anemones cluster on the whelk shell inhabited by a Hermit crab. When the crab is feeding the anemones lean over and sweep up any scraps it leaves. In return the Hermit crab receives protection from enemies through the stinging cells in the anemones' tentacles. Some Hermit crabs pick Sea anemones from rocks and transfer them to their shells and, when they move into a new shell, they take their anemones with them.

Chapter Ten
Social Lives

During the breeding season animals come together for the purpose of reproduction. Males must meet females, and among birds and mammals one or both parents stay with their offspring to fend for them. Groups of animals may also gather to make breeding a social event. Flocks of birds, colonies of seals and herds of deer are instances of communal breeding. The interactions between animals living together during the breeding season have been described in Chapters 5-7. Here we are more concerned with social behaviour not connected, or at least not directly connected, with breeding, as well as with the interactions between two species, ranging from co-operation to parasitism, which is the exploitation of one species' economy by another species.

In recent years zoologists have turned increasingly to the study of animal societies. In the context of breeding, they study not an individual or a pair but the relations between many individuals or pairs. Outside the breeding season, it has been shown that throughout the animal kingdom animals live as members of a community rather than as individuals insulated from their fellows. Communities are sometimes described as 'super-organisms' as a community is

Social weaverbirds, of south-west Africa, are famous for their massive communal nests. First a roof of grass is built, then the separate pairs build their individual nests under cover of the roof. These communal 'tenements' may contain a hundred or more nests.

218

Monkey recognizes monkey by face

Monkeys have been shown under test to be interested in pictures. Usually this interest wanes over a period of about one minute. A couple of years ago a scientist devised an apparatus by which each of a series of pictures would stay on as long as a button was held pressed. When he let his experimental monkey loose on the machine, the animal pressed the button and then pressed it again as his interest in the first picture waned. The new picture stimulated his interest once more. So the scientist could tell how much interest each picture aroused in the monkey by the amount of time the monkey spent scrutiniizing the pictures.

When a series of photographs of different monkeys was available, the monkey showed sustained interest in each, just as we would looking at photographs in the family album, and for the same reason, that we recognize each of the subjects as a distinct individual.

When the monkey pressed the button and saw one pig after another it quickly lost interest after the first one. Although the photographs were all of different pigs, they were all just pig to the monkey.

an entity like an organism, functioning as a unit with interdependent parts. Communities even have the power of healing damage like an organism. If a community suffers heavy mortality, increased births rapidly make up the loss.

Knowledge of community relations has revolutionized our appreciation of other aspects of animal life such as breeding, feeding and defence against enemies. The complexity of some animal communities is surprising. An example will demonstrate this. We have already met the idea of a hierarchy of dominant and subordinate animals, when discussing the selection of mates in Chapter 6. The degree of dominance or subordination is usually fairly simple. However, wild turkeys living in the Welder Refuge in Texas have most elaborate social strata. As with all discoveries relating to social organization the turkey's community life has been unravelled by following, over a period of several years, the careers of individuals marked with identifying tags.

In late autumn family parties of young turkeys break up, but instead of going their individual ways, young males stay with their brothers as a compact unit, and several of these units gather into flocks composed entirely of young males. It is at this time that the dominance hierarchy is established. This is done at two levels. Within each unit, the brothers fight to decide their ranking. Fights between two turkeys last for some hours, until the contestants are exhausted, and a dominant bird eventually emerges. A series of fights determines the position of each bird in the flock and, thereafter, each bird's position is assured for life. At the same time, the sibling units have communal battles with each other, and again the overall winners establish a clear and long-lasting dominance. Meanwhile, the females have their own flocks in which individual rank is disputed by the same process as in the males. Once ranks have been settled by means of quite vicious fighting, the turkey groups have formed a very stable society in which each individual knows his or her own place and fighting is very rare.

As spring approaches, the flock of adult males, which remains apart from the squabbling juveniles, splits into its own sibling units which go to the traditional mating grounds where they meet groups of females numbering about fifty hens. Only the dominant sibling unit moves about among the females; the others wait around the borders of the mating ground. The dominant unit displays to the females as a unit but only the dominant male within it mates, and he mates with all the females. Thus one, or perhaps two, males out of about thirty are involved in mating. After mating the females depart to nest, but the juvenile sibling units arrive to court any unattended females and some of the young birds enter the adult male flock, which has been thinned by a heavy mortality due to a variety of causes.

This is a complicated way of organizing the sort of lek breeding behaviour described for ruffs and Prairie chickens in Chapter 6. Surprisingly, not all turkeys behave in the same way. Turkeys of the eastern United States live in smaller, less rigid groups. The reason for the difference in social structure appears to lie in the ecology of the different populations. The Welder Refuge turkeys live in open dry grassland, where a sudden rainfall brings a bloom of vegetation. The multiple lek system is a protection against predation in open country, and may bring the females quickly into breeding condition when there is a sudden abundance of food. Eastern turkeys, living in woodlands with a more continuous supply of food, elect for polygynous breeding units of one male consorting with a harem of females. It seems that birds can make the most of a grassland habitat, with scattered rainfall and seasonally abundant food, by living in large groups. It may also be that in dense cover large groups have difficulty keeping in contact. This has been demonstrated for African weaverbirds, and is shown by the more solitary woodland Ruffed and Spruce grouse compared with the flocking Prairie chickens and Sage grouse of grasslands. In Africa prides of lions are larger on grassland than in thick bush.

The last point is important. We can no longer talk simply about the behaviour of a species as a fixed pattern of action common to all individuals of the species. A species' system of behaviour is as dependent on the environment as is an individual's response to immediate conditions of food supply and shelter. The social life of the turkey species depends on whether it lives in grassland or in woods. The same has been found to be true of baboons and chimpanzees. It is also true of man, whose cultural groups have embraced hunting or agriculture, settlement or nomadism, according to circumstances.

The role of the environment in moulding the communal life of a species, or population within a species, starts near the bottom of the scale of animal life. Many of the 'lower' animals form aggregations, perhaps around a rich source of food or in a sheltered place, but we can hardly call them societies. There may be some benefit to the individuals,

Mobbing and harassing
A crowd of small birds were gathered near a clump of evergreen shrubs, twittering excitedly. They did this every evening, and after a while, flew into the shrubs in small groups to roost. This evening there was a difference. The birds' chorus was louder, more excited, and they spent much time flying up into a tree and down again.
An owl was perched in the tree and the small birds were mobbing it. Finally the owl could stand their chatter no longer and flew off. After a while the twittering chorus subsided and the birds flew into the evergreens for the night.
In North America this would have been called harassing. In the United Kingdom it is referred to as mobbing. There is, in fact, room for the two words because the behaviour of songbirds towards their traditional enemies, the owls and the hawks, is broadly of two kinds, although the two tend at times to overlap.
In typical mobbing the central subject is usually an owl, which remains perched while the songbirds crowd around it, sometimes by the score, uttering their alarm notes. The owl is to all appearances undisturbed by the cacophony, yet it makes no attempt to hit back and ends by flying away followed by the more belligerent of its tormentors.
Hawks are 'seen off' by a crowd of small birds twittering and fluttering, or actively flying at them in a bunch. Harassing is the more appropriate term here, especially since it is not uncommon for one of the small birds to land on the raptor's back, causing it to spin to earth as it seeks to shake off its uninvited passenger.

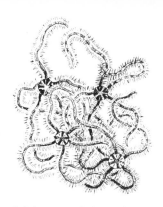

Brittlestars, relatives of star-fishes, may be found singly or in groups but some species form aggregations in which scores of individuals entwine their serpentine arms, clinging bunched together.

however, although it may be intangible. For some reason cockroaches perform better at learning tests when in small groups than when alone and, as aquarists know, goldfish grow better when kept together, probably because of a pheromone secreted into the water. Around the eastern shores of the United States there is a brittlestar, a sort of starfish, which lives among eelgrass and spends much of its time with its arms entwined around the grass stems. In winter, the eelgrass dies back and the brittlestars cling to each other, to form closely packed bunches. Experimentally, it has been found that bunched brittlestars live longer than solitary ones, but the reason for this is not known.

Aggregations of fishes in shoals and birds in flocks confer several advantages, both in bringing individuals together for breeding and for utilizing food sources. They also act as a defence against predators. When confronted with a compact mass of animals, a predator is confused and finds it difficult to pick out one quarry on which to concentrate its attacks. Gulls, terns and starlings come together when a falcon is sighted, and fly erratically to and fro in a tight mass. The falcon's attack is foiled unless it can snatch a straggler. Some fishes behave in the same fashion. Thus, goldfish find it difficult to feed on dense masses of Water fleas (*Daphnia*). Each time a goldfish goes to snap a Water flea its attention is diverted by others nearby. These aspects of flocking and shoaling are discussed in more detail later in this chapter.

Social insects

Outside human societies, the most complex social life is found in the Hymenoptera, the order containing the social insects, and the Isoptera, the termites. A society implies a stable relationship between individuals of an order greater than that between sexual partners or gregarious groups. The best known, and most important, of the social insects is the **honeybee**. Of the nineteen thousand species of bee only a few are social; the majority lead solitary lives. There are three species of honeybee, of which *Apis mellifera* is the most widespread. The success of the honeybees lies in an ability similar to that of mankind. The bees can control their environment, regulating the temperature and humidity of the hive and surviving the winter on stored food. The bee society is based on a system of **castes**, giving a **division of labour**. At the head is the queen, one to each colony, a fertile female who becomes an egg-laying machine. Fertile males are called drones. They develop from unfertilized eggs and their only function is to fertilize the young virgin queens when they leave the colony on a nuptial flight. The majority of the population are workers, infertile or sterile females which develop from fertilized eggs. At the peak of production, the queen bee lays fifteen hundred eggs per day and the colony may number sixty thousand workers.

The nest of a honeybee colony consists of a series of combs made of wax secreted from the workers' bodies. The eggs are laid in the hexagonal cells of the comb, one to each cell, and the larvae complete their development within the cells while being fed on pollen and honey by the workers. After they have emerged from the cells, young workers follow a stereotyped working life. Their first chore is to clean the hive. When a few days old they develop protein-secreting glands with which to feed the larvae. Shortly afterwards, their wax glands start to secrete and they build and repair the comb. During this period young workers process nectar into honey and store it, with the pollen, in parts of the comb. The atmosphere in the nest is also controlled by some of the workers spreading water on the comb and fanning with the wings to make it evaporate. The precise duties within the colony depend on the needs of the colony as a whole at the time.

When two or three weeks old, the worker is ready to leave the nest to collect food. The transition is gradual. First, the worker acts as a guard at the hive entrance; then it makes short flights to learn the landmarks which will later guide it home. Eventually it is ready for foraging. For four to six weeks, until it is worn out and dies, the worker visits flowers to collect pollen and nectar. Water and resin from buds are also collected. The resin is turned into propolis, a sort of cement used for filling crevices in the nest. Bees are attracted to flowers by both smell and sight, and many flowers have a **nectar guide**, a central 'target' of lines on the petals that converge on the site of the nectary at the base of the flowers. Nectar is carried in the crop where it mixes with enzymes that convert the glucose into two other sugars — laevulose and dextrose. Excess water is later removed from this mixture to make honey. Pollen is picked up on the hairy body and is combed off by the legs, moistened with a little honey and stored in the pollen baskets or corbicula on the hind legs, where it can be seen as two yellow balls. The food brought back is passed from mouth to mouth among the workers. Food-sharing, or **trophallaxis** appears to be the bond which holds the colony together.

Ant farmers

Long before history, man tamed and domesticated certain animals for his own uses, more especially for food. Ants have been doing much the same thing with aphids for infinitely longer, in order to eat their honeydew and take it back to their nest for the larvae. Just as man is able to stimulate the production of milk in cows and goats, so too can ants encourage production of honeydew in aphids by improving their living conditions. They may 'herd' their charges, by forcibly confining them to the growing tips of plants which are the most nourishing. When there are no ants present the honeydew may cover such large areas of the plant that it eventually wilts and dies. By removing the honeydew the ants ensure their charges' food sources.

The ants may also take aphids into their nests where they lay eggs or they may carry the eggs themselves from the plant where they are laid. After hatching, the young aphids are carefully tended and 'milked' by the ants while they feed on the roots of various plants. Some aphids live only in ants' nests while for others the ants build special shelters where they can feed, protected from predators. Comparison with the human farmer's cattle-sheds is irresistible. Honeydew is not produced continuously and the drop of liquid produced is in normal circumstances discarded by a flick of the aphids' hind leg. Under stimulation from an ant, however, the aphid does not discard the fluid but allows the ant to remove it, and goes on doing so, seeming to enjoy the caressings of the ant's antennae.

Under continued stimula-

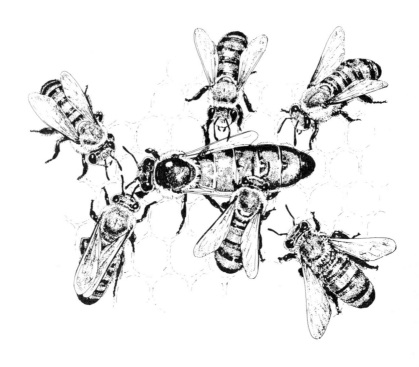

tion very large amounts of honeydew may be produced. One large aphid can produce nearly 2mm³ in an hour, and a colony of the Common ant *Lasius fuliginosus* can, it has been estimated, collect about 3-6 lb of honeydew in 100 days.

A colony of honeybees is a miracle of organization. The individual bees appear very purposeful in carrying out their manifold tasks, although there is no central 'intelligence' guiding their actions. Workers carry out particular tasks as the need arises. If food is short, for instance, workers start to forage at an earlier age. The overall control of the nest's activities comes from pheromones secreted by the queen. Most of these are produced in glands near her mouth. Together they are called 'queen substance'. The queen spreads them over her body and they are licked off by the workers, who pass them to the rest of the colony by trophallaxis. Queen substance prevents the workers from rearing new queens; if the queen dies, the lack of queen substance in circulation encourages the production of a replacement. Another pheromone is secreted by guard bees to alert other guards when danger threatens and they distribute a unique 'colony odour' to help guide returning foragers. Stranger bees do not have the colony odour and are attacked.

At the height of the summer the nest becomes very crowded. The spread of queen substance is disrupted and new queens develop. They leave the colony, mate with the drones and, collecting a band of workers, fly off in a **swarm**.

223

The swarm settles on a tree and scouts are sent out to find a new home, a hollow tree for instance, unless they are captured by a bee keeper and transferred to a hive. At the end of the summer most of the workers die, but some build up fat reserves and survive the winter.

Other **social bees**, such as the bumblebees and the stingless bees of the American tropics cannot maintain the colony over winter. Only the young queens of these species survive. In the spring, they awake from dormancy and found a new colony, collecting food and caring for the larvae until there are sufficient workers to take over the chores. The same is true for the **social wasps**, some of which are called hornets and yellowjackets. The wasps have a more liberal diet. They feed on many sweet liquids, to the consternation of picnickers, and catch other insects to suck their juices. A wasp's nest is made of 'paper'. Wasps can be seen, and heard, rasping at the bare wood of dead trees and fence posts. They mix the wood with saliva, carrying away balls of pulp which are used to make the cells and walls of the nest.

Whereas there are solitary bees and wasps, all the thirty-five hundred species of **ants** are social. Their social life is similar to that of bees, although after mating the queen sheds her wings, and the workers are wingless. In some species, there is a special caste — the soldiers, distinguished by large heads and strong jaws. Soldiers are particularly numerous in the colonies of Army ants, which are also known as Driver or Legionary ants. These ants are nomads; their moving columns are a scourge in tropical countries as the ants are carnivorous, and a mass of them can devour animals many times the size of an individual ant. Other ants are vegetarian, such as the Leafcutter ants which strip leaves from trees and use them as a basis for growing fungi on which the larvae are fed. Harvester ants live on seeds which they lay out to dry and nip to prevent germination.

The **termites** have evolved a way of life similar to that of the hymenopterans but the male, or king, continues to live in the nest with the queen after mating. The queen's abdomen swells enormously in pregnancy and she becomes an egg-laying factory, producing an egg every two seconds; and a queen termite has been known to lay eggs for fifty years. Except when swarming, most termites live underground. Their principal food is wood, which they digest by means of protozoans living in the intestine, except in one family which harvest bacteria and fungi specially grown on collected wood fragments.

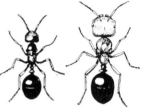

Castes of Harvester ant Nessor barbarus: *winged queen and male (above, left and right), wingless worker and soldier (below, left and right).*

Spawning crisis
The lower invertebrates, the lowest form of animal life, bear some comparison with Shakespeare's lean and slippered pantaloon 'sans eyes, sans nose, sans teeth, sans everything.' Such sense organs as they have are few and elementary. Their nervous system is of simple design or non-existent, and a brain is wholly lacking. Some do not even have the means of locomotion and in those that have movement it is slow and restricted. Most do not even come together for mating but simply shed their genital products, the ova and sperms, into the sea and let providence do the rest.

Yet although they appear to live in such splendid isolation they are not wholly without means of communication, although they do not have even the beginnings of a language. Such lowly, seemingly isolated animals can be epitomized by the Sea urchins, prickly relatives of starfishes. There is one great occasion in the lives of Sea urchins when communication is vital. This is when they are approaching the time for spawning. If the members of a Sea urchin community did not spawn simultaneously the chances are that a high proportion of their eggs would not be fertilized. This is obviated by the spawning crisis. The first urchin to liberate genital products into the sea also liberates a pheromone. This stimulates the nearest urchins to spawn and spawning rapidly spreads throughout the area.

Shoaling mackerel keep station with each other by sight and by picking up vibrations in the water from movements of their fellows.

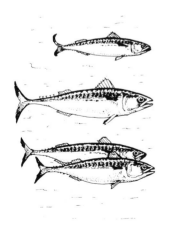

Shoaling of fishes

The fishing industry is based on the shoaling or schooling habits of various fishes. A shoal of fishes is more than an aggregation of a number of individuals. It is a structured entity in which there is a co-ordination of behaviour. If, for instance, some fishes are disturbed — a stone lobbed into their midst does the trick — a mere aggregation disperses in panic, but a shoal moves as a unit, each individual keeping station on its neighbours. Keeping station is achieved partly by vision and partly by the vibrations from neighbouring fishes, detected by the **lateral line** system (see p 40). A blind fish will not shoal and, if blind in one eye only, a fish will station itself in the shoal so that other fishes are on its 'good' side.

The size of a shoal ranges from about two dozen fishes or less in a shoal of Bluefin tunny, to millions in a shoal of herrings. Herrings, tunny and mackerel, for instance, are lifelong shoalers, but sticklebacks shoal only outside the breeding season.

The development of shoaling has been studied in young whitebait. After hatching, they are solitary, and if two accidentally meet they will flee each other. As they grow, however, one whitebait will approach another from the rear and they will swim together for a while. Swimming in tandem gradually becomes more common and more fishes are added to the line. Station-keeping becomes more orderly and, by the time the young fishes have doubled or trebled their original length, they are forming well-ordered shoals.

Shoaling is a feature of both plankton-eating and carnivorous fishes. The menhaden is a herring-like fish of the eastern seaboard of the United States, south of Cape Cod where the range of the herring ends. It lives in enormous shoals which, like other members of the herring family, feed on plankton. They swim with their mouths open and trap plankton from water passing through the gills by means of a sieve of intermeshing rods called gill rakers. The shoal swims to and fro, occasionally breaking the surface and even leaping clear of the water. It may swim in circles 'like the dust driven by the whirlwind'. Menhaden are the food of many animals, including dolphins and porpoises, bass, tuna and bluefish. The bluefish is a shoaling carnivore. Its shoals, again numbering millions, will suddenly appear in an area. Bluefish will slaughter other kinds of fish, behaving like 'animated chopping machines' and leaving a trail of bloodstained water and fragments of flesh.

One of the benefits of living in shoals is, as has been mentioned, that it confuses predators. Not only does a predator find it difficult to single out one animal from a group, but the effects of predation are, surprisingly, less if animals bunch together. The number of fish such as herring or menhaden that a predator can consume is naturally limited by the predator's hunger and catching ability. If a shoal exceeds this number the chance of an encounter between the predator and any one fish is reduced.

The social life of birds

Flocks are largely similar in structure and function to the shoals of fishes. How a flock of hundreds or thousands of birds can manoeuvre without collision is an unsolved problem, but it must be assumed that each bird is keeping a very close watch on its neighbours, and that the marvellously precise wheeling of a flock of queleas, one of the Weaver birds, of Africa, is controlled by intention movements. By noting the 'intention' to change course by one of the leading birds, the followers are ready to change direction to suit. Intention movements are also likely to be the signal for a flock to take off in unison. One bird flexes its legs and starts to open its wings in preparation for springing into the air. Others notice this and in a split second all are ready for take-off. Cohesion of the flock is improved by 'flock markers' in the form of flashes of white feathers on wings, tail or rump. These are particularly common in waders. Special flight calls also help to keep the flock together, especially in cover.

The precise functions of bird flocks is a matter of some debate, but it is possible that functions vary from species to species. Outside the breeding season there are two advantages of flocking: to escape predation and to increase the chances of finding food. The escape from predation has been mentioned already. A tight bunch of starlings flying erratically discourages the swooping falcon. The predator finds it difficult to single out its quarry and, furthermore, it is possible that it faces the danger of injury through collision (see Chapter 4). As with fish shoals, flocking may also reduce the chances of a single individual being attacked and it has been suggested that by living in flocks rather than being spread evenly over the ground birds reduce the chances of being found by the hunting predator. A similar advantage has been suggested for birds that nest in colonies. A hunting fox, for example, is more likely to stumble on a single bird's nest if the nests are well spread out, but it may completely

Opportunist fish
It happens many times in the world of animals that two of them live together to the advantage of one or both. This is given the omnibus title of **symbiosis**. If the advantage is one-sided it is called **parasitism**. When two such partners share the advantages, especially when this consists of sharing food it is called **commensalism**, or messmates.

In the warm seas of the Caribbean and the southeastern United States live two fishes, a Puffer fish and a scat. The scat has a weak mouth and jaws, and can eat only soft foods. The Puffer fish, so called because it inflates its body — puffs itself up — to make it harder for a larger fish to eat it, has jaws and teeth like a parrot's beak, and just as strong.

When a Puffer fish has caught a shrimp it crunches it up — and drops crumbs of flesh. The scat hangs around below the Puffer and catches the crumbs.

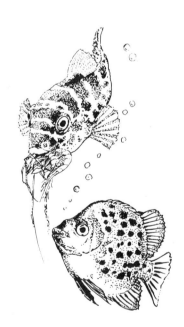

226

When a sparrowhawk flies into a given air space the immediate reaction of starlings is to bunch in a tight 'box' and to fly parallel to the course of the hawk, at some distance from it, but near enough to constitute a threat.

miss a colony if it happens not to venture into that particular area. On the other hand, once it does find a colony it can wreak havoc in a short time.

Flocking also increases the chances of an individual being alerted to danger. A solitary bird has frequently to break off feeding to scan its surroundings for a stooping falcon or a prowling cat but if it is a member of a flock it can spend more time feeding, because there will always be a few of its fellows on the alert that will give the alarm. The extra pairs of eyes that help keep a watch for predators also help in the search for food. In the woodlands of Europe, after the breeding season has finished, small flocks of tits can be seen making their way through the twigs and foliage. Their food, insects and other small animals hidden in bark crevices or among the leaves, is distributed patchily. As the birds move about, keeping in touch with shrill calls, they keep an eye on each other. When one tit finds an insect, the others come over to investigate and then start to search the same kind of place. Thus, if there is a particular kind of animal in abundance on a certain tree, the flock soon learns to exploit it. When the supply is used up, the birds move on and start to look for other food.

Compared with woodlands, grasslands provide a more regular habitat in which small animals are evenly distributed. Nevertheless, feeding in a flock still confers an advantage on the individual because less time is spent searching for food. The Cattle egret, a member of the heron family which has spread from Africa into Australasia and America within the last two centuries, feeds on insects that it finds in open grasslands. Flocks of Cattle egrets sweep forwards in a leapfrogging motion. As individual birds find themselves in the rear of the flock they take off and fly forwards, landing in the vanguard. In this way each Cattle egret takes its turn at the front, where there is an abundance of insects. Its movements also stir up these insects so that they are readily seen and caught by the birds coming up behind.

Cattle egrets have learned to use other agencies to do the stirring up. They follow cattle (hence their name) as well as antelope, elephants and even motor vehicles. It has been found that Cattle egrets catch one and a half times the number of insects, for less expenditure of energy, when following cattle than when feeding on their own. The American ani, a bird of the cuckoo family, engages in the same trick. It has been observed that an ani can catch three times as many insects if it follows a cow. On the other hand, anis do not feed in a flock like the Cattle egrets, so it is strange that they should make use of other species in finding food when they do not help each other.

Relationships within a bird flock vary from complete anonymity to recognition of each individual's status by the other birds. In the large flocks of waders or shore-birds that gather on the seashores, in the vast flocks of starlings that fly into city roosts each evening, each bird does not know its neighbours. The relationship between the individuals is a finely adjusted balance between attraction and repulsion. The birds are attracted to form a flock, but there is an antagonism that keeps them spaced. Each individual has sufficient room to feed and to take off without hindrance, because the result between the opposing forces is an even spacing between the birds. The distance between each bird is called the **individual distance**; it varies from species to species and within a species may change with circumstances. If one bird encroaches on another's individual distance the latter will either move away or drive the interloper back.

Where a flock is more stable, individuals learn each other's identity and relationships develop between them. Each recognizes the social status of other members of the

Natter-buddies

Starlings are natter-buddies. This is true of the whole family, which includes the mynahs, famous for their ability at mimicking sounds, including human speech. All starlings are fully deserving of the name 'natter-buddy' because they, like human natter-buddies chatter away without listening to what the other is saying. The Common starling reserves its nattering especially for evenings just before going to roost.

As the sun goes down the starlings gather in a noisy crowd in the top of a tall tree and the chorus builds up as other starlings fly in. As the crowd swells, so does the chorus.

There is a group-word for the vocalisations of starlings, it is murmuration. But this evening chorus is not murmuration. It is sheer nattering, and it seems to provide a focal point directing other starlings in the neighbourhood to the assembly point.

The strangest feature of it is that suddenly, as if at the final sweep of a conductor's baton, all fall silent. The sound stops as if cut off by a knife. A second later, the whole flock takes wing, as if on a word of command, and all head for their roosting place, often the city buildings several miles away.

This sudden cessation of sound is most baffling. If the birds only performed nearer the ground level one might be able by careful observation to unravel the mechanics of it. There must be some signal for it either from the environment or from a leader, but although it is difficult to visualise what it might be, there must be some form of communication or understanding between the starlings in a flock to bring about such a striking cessation in such a loquacious group.

flock and a **peck order** or **dominance hierarchy** develops. The expression 'peck order' was coined during studies of relationships in a flock of domestic chickens, when it was found that there was a definite order of precedence at feeding time. Precedence was established by fights when the chickens were first put together and was enforced by chickens being pecked by their superiors when they forgot their social standing. At the head of the peck order or hierarchy is the dominant Chicken A which takes precedence over and can peck any other chicken. Chicken B may peck all others except Chicken A, Chicken C may peck all except A and B, and so on down to the subordinate chicken at the bottom which is pecked by all the rest. The peck order brings stability and peace to a flock because each member knows his or her own place, and once there has been a series of fights to determine position, discipline is upheld by a quick peck at any that step out of line. Provided the dominant bird can uphold its authority, life is peaceful, except for the bird at the bottom who may die through nervous strain or starvation. Fights break out when the dominant bird is seriously challenged or when new birds join the flock.

The peck order of chickens is a simple system. More complex hierarchies are found in many other animals, not only in birds but in lizards and in many mammals. For instance, as we know, Konrad Lorenz found that jackdaws mate for life and that the female takes her husband's rank. The complexity of wild turkey societies has already been described; mammal societies are discussed below. The peck order of chickens is called a **linear hierarchy**. Other hierarchies include triangular relationships (A pecks B, which pecks C, but C pecks A) or the ganging up by two subordinates to outrank an animal that would dominate them individually, as happens in baboons.

Friendly activities

In some birds and mammal societies, the mutual repulsion between individuals is removed and individual distance becomes zero. In other words, the animals are drawn together. Male Emperor penguins, for example, huddle together while incubating their eggs in the rigours of the Antarctic winter. Each bird rests its bill on the bird in front and by pressing together they keep each other warm. For other species **clumping**, as this is called, appears to be merely social; the birds seem to derive a sense of security from the touch of another. Lovebirds, small African parrots popular

African lovebirds are small parrots that have the habit of huddling in pairs or in clumps, shoulder to shoulder, beak to beak, frequently indulging in mutual preening. It is said that if one partner of a pair dies the other pines and finally dies of grief.

with bird fanciers, are named for their 'loving' habit of huddling together. Avadavats, small red and black cagebirds from India and Southeast Asia, spend much of their time in clumps, leaning against each other so firmly that the outside birds may have to prop themselves with a wing.

While clumping, or at other times, these socially inclined birds may preen each other. This is called **allopreening** (which is to be distinguished from **autopreening,** when a bird preens its own body) and it is also part of the courtship behaviour of many birds. The actions are the same in both allopreening and autopreening except that a bird usually allopreens only its partner's head and neck. An avadavat will solicit allopreening by 'offering' its head with feathers ruffled, and it does not merely want an awkward part of its body cleaned. 'Dirtying' avadavats with flour does not increase the amount of allopreening. The action is a sign of goodwill between the two birds and helps to reduce aggression.

In mammals, the equivalent to allopreening is called **allogrooming**. Mutual licking is a common sight among domestic cats that share a house and the grooming and cleaning by monkeys of each other's fur can be seen by anyone visiting the monkey house of a zoo. Apart from the actual cleaning process, removing flakes of dead skin and clearing tangles, allogrooming seems to be a pleasurable activity. It is also used, like allopreening, to foster friendship. In baboon troops, the dominant male is groomed by his subordinates, apparently as a means of promoting 'togetherness' and reducing animosity.

Social life of mammals

In mammals, as with birds, there is a great variety of social organization. There are solitary mammals that come together only for breeding purposes, and mammals that live in huge herds, as did the bison, springbok and zebra before they were severely reduced by man's gross overexploitation. The mammals that live in herds enjoy the same advantages as flocking birds. The herds tend to keep predators at bay and may, to some extent, make efficient use of the available food. The close study that has been made of mammal societies in recent years shows that there is an infra-structure within the large aggregations or herds. The members of a herd, troop, school or other group of mammals are not equals, as in a shoal of fishes, and there is often more than a simple peck order or dominance as in many birds. There may

Hunting dogs talk
Domestic dogs in Australia sometimes go feral and form up into packs. In reply to an enquiry about the dogs' behaviour the following letter arrived.
'I have never seen a pack at work but have seen them returning from a hunt and on several occasions have seen dogs gather into a pack. On each occasion there was at least one fox terrier present.
'I can give you a very accurate report of the behaviour of a rabbiting pack. When we bought this property of 4000 acres about nine years ago, it was one huge warren. We actually skinned 37,000 rabbits and poisoned, shot, trapped and clubbed as many more.
'We had six descendants from a wire-haired terrier. These were bred from all kinds of mongrels and one pure bred German collie. All of them were four or five generations away from the terrier. They made an amazingly efficient pack.
'Scouting ahead of us on a front of from one to three hundred yards, depending on the height of the grass, they seldom passed a rabbit, frequently standing on the hind legs to check position in relation to the other dogs so as to see if one was setting a rabbit. If one set a rabbit the remainder raced to the spot and surrounded the area. Not one rabbit in fifty would escape. They used different barks to indicate various activities such as: I am chasing a rabbit or I have one in a log, or I have found a burrow, or I have a cat or a hare or fox or kangaroo. My sons and I recognized most of these barks. One of the dogs, our beloved Bill, could tell, without fail, if there was a rabbit in a log or burrow. If he sniffed and walked away the others at once lost interest and continued their search.'

be a quite complex society of sub-units, based on social status and relationships. But to discriminate between bird and mammal societies may be artificial, as is suggested by the newly-discovered complexity of the social life of wild turkeys, already referred to. To date, bird societies have not received the same close attention as those of mammals, although it does seem that, in general, mammal societies are more stable and long-lasting than those of birds.

Once again, it is the marking and tracking of individuals that has enabled us to recognize social relationships. A House mouse, found stealing food in a larder or cowering in a trap, may seem to be a solitary animal, but every mouse is part of a community based on a home range. The leader of the community is a dominant male who has exclusive access to the breeding females. Young animals are tolerated. Young females are eventually recruited as breeding animals, while young males must either seek their fortunes elsewhere or remain as subordinates, physically and socially immature, until the dominant animal dies, when one of them will emerge as the new overlord.

Tracking animals day-by-day over a long period gave J.H. Kaufmann the chance to learn about the private life of the coati. The coatis became so used to his presence that he could follow them as they foraged in bands through the forest for insects and other small animals. The band keeps together through chittering calls and there is little aggression between the coatis. At dusk they retire to a favoured tree where they have nests of twigs that may be shared by several coatis. These bands consist solely of females and their young. Each band has a home range but this overlaps neighbouring ranges. Males lead solitary lives, but during the breeding season one is allowed into each band.

A frequent sub-unit of mammal societies is the **bachelor group** of young males that live separately from the adults and take no part in breeding. Bachelor groups exist more particularly in polygynous societies where older males gather harems of females or, at least, dominate the mating grounds. If there is a sex ratio of one to one, it follows that there will be a proportion of males excluded from mating. These form the reserve from which recruits are drawn for breeding. For example, the birds on the edge of a blackcock or Prairie chicken lek move to the centre as older birds die. Bachelor herds tend to occupy poor feeding grounds, as in the vicuña, where the dominant males occupy territories in which they keep a number of females and their young on the best

231

pastures, and the bachelors roam on the fringe areas. Young males join bachelor groups when they are driven from the breeding groups by the adult males, or they may form groups of their own accord, as do zebras. The basic social unit of Burchell's zebra consists of a stallion with up to half a dozen mares and their foals. Movements of the unit are usually led by a dominant mare, with the others following according to seniority. The stallion may occasionally lead but usually he brings up the rear. This unit is very stable, and continues outside the breeding season, unlike many other harems. The stallion is eventually replaced because of sickness or death; otherwise the young stallions have to abduct young mares to set up their own harem. A similar situation is found in domestic horses, but the behaviour of the stallions varies with the breed. Some lead the mares out of danger while others bring up the rear, sheltering them from attack from behind.

Mountain sheep, such as Dall's sheep or the Bighorn sheep of North America, have social systems in which the horns of the rams play an important part. They fight throughout the year as a way of establishing dominance. In their head-on

Bighorn rams fighting to establish dominance in the hierarchy of the herd. This involves a tremendous clash of foreheads and horns, the horns sometimes being damaged in the encounter. To some extent also the horns are a badge of rank: two rams meeting turn their heads to show off their horns to each other, the one with the smaller horns then retiring.

Sloth mystery solved

It has long been known that sloths have green algae living in their hair. It has also been known for a long time that sloths have moths in their hair. There knowledge ended and mystery began, because there are three male moths to every female and there are no eggs, larvae or pupae to be found among sloths' hair.

The mystery appears to have been set at rest this year by two zoologists working in the Panama Canal Zone. They kept the female moths in the laboratory to lay eggs. The larvae hatching from these were offered sloth hair. They would not eat it. They were offered leaves of the kind sloths eat. They ignored them. When offered sloth dung they made a hearty meal.

Then the rest of the story was unravelled.

A sloth defecates once a week, more or less. It descends from the tree tops to hang from a vine by its forelegs, so that its hind feet touch the ground. It scratches a hole in the ground and the female moths go down into it to lay their eggs on the dung. Slow though the sloth is, most of the females miss the return journey, hence the preponderance of males on the sloth's coat. The larvae feed on the dung and pupate in the hole. The adult moths emerging from the pupae fly up in search of sloths in the tree-tops.

One sloth tracked by a radio collar was overtaken by an average of one moth a day. This is **symbiosis** (see p 237) brought to a fine art!

clashes, the attacker attempts to deliver a cutting blow with the edge of one horn. His opponent parries it with his head, which is protected by a double layer of bone as well as the coiled horns. Shortly after leaving their mothers, young rams join male-only groups. They are protected from the attacks of their elders by acting in a feminine manner. The hierarchy in the group is based on horn size and the young rams follow the largest-horned ram. In these groups the young rams learn their way around the tortuous mountain paths to the scattered pastures. Only the dominant ram breeds; it joins the group of ewes on its separate range. The ram does not defend a harem but guards each ewe as she comes into season.

Most of the hoofed animals live in herds structured along the lines described for zebra and Mountain sheep. A similar set-up occurs in Sperm whales, where there are separate schools of young bulls and of cows with immatures. Mature bulls are solitary, except when they join the cow schools in the mating season. The bulls fight for the privilege, with flesh wounds and even broken jaws befalling the losers. Hunting mammals usually lead a more solitary life, but the dog family contains many exceptions. Wolves hunt in family parties and Hunting dogs run down large hoofed animals in packs. Hunting dogs defend a territory around the burrows where both males and females guard the young and a larger range is used for hunting.

The lion is unusual in being a social cat. The pride consists of a number of lionesses with their cubs and immatures. Trespassers into the territory are attacked. The male lions associated with the pride are autocrats (see p 62). They take little part in hunting; indeed they are poor hunters, as they are less agile than the lionesses and their manes make them conspicuous — like 'animated haystacks' as George Schaller has described them. Yet the lion takes precedence at the kill, literally enjoying the 'lion's share' of the food. Other cats 'walk by themselves'. As can be seen, domestic cats have ranges of interlocking paths and have a loose system of rank which is decided by fighting. However, cats have a social gathering at night when they meet at a particular place and merely sit around. It is a friendly meeting and the cats often groom each other.

Societies of monkeys and apes

The family and social lives of primates have come under intense scrutiny because they are our nearest relatives, and the organization of their societies may shed light on the origin and evolution of human family life and group organization. By understanding our social origins we may perhaps be able to understand present difficulties such as the breakdown of family life. And with understanding, there is the greater possibility of cure. Unfortunately, the situation is not so simple. We are not chimpanzees, nor baboons nor leaf-eating langurs. We are human beings, a species separate from other primates adapted to a different way of life. Therefore, it is to be expected that we should have a different social organization, and that as we live in widely varied habitats, the organization will vary from place to place. The African bushmen are very different from the tribesmen of the New Guinea highlands and they, in turn, bear little resemblance to the Eskimos. Nevertheless, it is still of great interest to compare the behaviour of man and other primates and, in view of the differences between human tribes, it is interesting to note that the social life of monkeys and apes also depends on environment.

The primates originated as tree-dwelling animals rather like present-day Tree shrews and, while many are still arboreal, there has been a movement by some, including ourselves, to 'come down from the trees' and live on the ground in the forest clearings or in open savannah country. Primitive primate societies may well have been like that of the modern lemurs of the Madagascan forests. They live in mixed troops of fifteen to twenty individuals, each troop having a territory the boundaries of which are marked with scent and defended by calling. The troops are stable units, with little fighting between or within them. There is a hierarchy in the troop, but the members are friendly and often groom each other. If this is the basic primate community, it is retained by some of the monkeys which still live in troops in the trees, in what we may think of as a typically 'monkey' existence. Forest-dwellers usually live in small groups because of the difficulty of keeping a large group together in dense cover, and the forest is the home of the very solitary orang-utan.

Altogether, life is freer in a forest. There is a regular abundance of food and less danger. Monkeys and apes can flee to the safety of the trees and there are not the big predators, such as lions and cheetahs. In open country, both competition for food and defence against predators necessitate a strict social organization. Chimpanzee societies are free and easy, and groups intermingle, while baboons on

Octopus burns its fingers

For the last hundred years every textbook and almost every children's book on animals has contained a picture of a Hermit crab with a Sea anemone seated on its shell. It is the stock example of what is called **commensalism** (p 237), living together for mutual advantage. Up to 1971 it was known that this is a partnership voluntarily undertaken, the anemone willingly crawling onto the Hermit's shell, the crab conniving to the same end. What was in doubt was the advantages, if any, gained by the two partners. Theories were that the anemone stung and paralyzed prey which the crab ate untidily, the anemone taking the crumbs; that the presence of the anemone disguised the crab's adopted shell, so hiding it from predators; and that the anemone drove away predators. Nobody was sure until the Canadian zoologist, D. M. Ross, brought an octopus and a Hermit crab with its anemone together in an aquarium. Octopuses eat crabs and Ross saw and filmed this octopus trying to capture the Hermit crab. It made one attempt but quickly let go with its arms and retreated. It tried again, less certainly, and again beat a retreat. This went on until the octopus not only left the commensals alone but fled before them.

As every schoolchild knows, the Hermit crab has a soft abdomen and, at an early age, looks around for a mollusc shell, in which to shelter this vulnerable part of itself. Moreover, as it grows in size, the Hermit must transfer to bigger and bigger shells. An octopus, seeking to eat a Hermit must insert the tip of one of its arms into the mollusc shell to winkle out the Hermit, and this may take hours, or even days.

So the Hermit not only adopts a highly protective armour (the mollusc shell) but, when it adopts an anemone as well, it achieves virtually 100% impregnability.

Granny monkeys
Hierarchy is a word that fifty years ago, or even twenty, was used only by students of sociology and politics. Times have changed, and even if we do not say the word the notion it represents is familiar — that in any group of animals there is a boss and the rest occupy positions of varying degrees of subordination.
There is a tendency to think of a hierarchy as a male prerogative, which is misguided, for there are female hierarchies. One of these that has been recently studied is found among female Hanuman langurs, and one of the more interesting aspects of it concerns the fate of old females. They go to the bottom of the scale automatically with increasing age. Moreover, while doing so their social functions change.
When the langur female is past her reproductive maximum her actions become altruistic. One of the first signs she has reached this stage is that she has second choice at the food after the young females. She becomes responsible for the defence of the troop, attacking predators, including man and dogs. She will ignore her own safety to rescue langur babies from aggressive males when their own mothers do nothing about it. Indeed, like the human granny there is altogether a greater element of altruism in her life than she showed as bride or mother.

the open savannah lead a regimented life. There is sexual dimorphism among baboons, which produces powerful males with heavy manes and long canine teeth. A strict dominance hierarchy gives the largest males access to the females. On the move, the troop assumes a certain order. In the centre are the dominant males, their consorting females in season, and the females with infants. Around this core are scattered the young and sub-adult animals. When danger threatens, in the form of a lion or cheetah, the dominant baboons move to place themselves between the danger and the troop, and may even attack. This system is used by the savannah-dwelling subspecies the Yellow baboon, while the Olive baboon of Uganda lives in woods, and is more arboreal and has a more easy-going life. The hierarchy is relaxed and there is no 'order of march', perhaps because safety is easily assured by climbing trees.

For those who want to compare man with his nearest relatives, it is instructive to note that there is little aggression between primate troops. This may be because there is plenty of room and they rarely meet, although they are often amicable when they do meet. There is also little fighting between individuals. Fighting does occur when a young male is establishing his place among the adults, or when the dominant animal is losing his grip and is being challenged. Several writers have pointed out that primates are aggressive when overcrowded in captivity, and overcrowding could be a major cause of human aggression. Another interesting point is that our early ancestors lost the large canines that are the weapons of other primate males. Instead, we have large brains, which could have developed as an alternative — the use of guile replacing physical weapons in the struggle against predators and rivals.

Altruistic behaviour
The defence of the baboon troop by the dominant males is a very special kind of behaviour. The best form of defence for any animal is retreat, but these baboons deliberately expose themselves to the danger of a violent death for the sake of their fellows. Such behaviour is **altruistic**: the community benefits at the expense of the individual. Altruistic behaviour is most often exhibited by parents in defence of their offspring. The alarm calls and distraction displays of birds described in Chapter 4 are altruistic, as is the swooping of Black-headed gulls on foxes and man. It would be much safer for the adult gulls to flee from such predators, but they

seek to distract them from their broods, by swooping down and then shearing away just above the predator's head. In general, altruistic behaviour is not overdone — the parent stops short of actual sacrifice. Death of the parent will not help the family, and from the point of view of the individual, it is better that a parent lose the family and live to breed another day.

Another form of altruistic behaviour is **epimeletic** behaviour, in which an animal comes to the rescue of a companion in distress. This most commonly takes the form of nurturing a wounded youngster. When a young baboon hurts itself, its cries bring nearby adults to its aid, and even adult baboons may receive solicitous attention when sick or wounded. However, epimeletic behaviour appears to be best developed in dolphins. In one instance, a captive Bottle-nosed dolphin was accidentally knocked out. Its two companions lifted it to the surface so that it could breathe. Similar behaviour has been recorded in the wild (see p 299).

Animal partnerships

The most obvious relationship between two species is that of predator and prey, or its modification of parasite and host, but there is a range of interspecific relationships which

Dolphins in distress
A distress call is a feature of the behaviour of social animals and there are few animals as sociable as dolphins. Occasionally a school of dolphins becomes stranded and attempts at rescue are made by towing them off the beach. Unfortunately, these attempts are frustrated when the liberated dolphins immediately cast themselves back on the beach. It seems that the distress calls of their stranded fellows act as an irresistible draw and in these circumstances the calls are worse than useless. However, this is not the case when the calls of wounded dolphins bring others to their aid. There are well documented stories of dolphins, both in the wild and in oceanaria, lifting wounded or unconscious companions to the surface.
Helping an injured companion is known as **epimeletic behaviour** and, in dolphins, is not re-

stricted to helping members of the same species. There are reports of dolphins helping drowning men and a sick Common dolphin emitted distress calls when an attempt was made to remove it from an oceanariam for treatment. Its companion, a False killer whale, pushed it away from its captors and when the dolphin was recaptured the whale seized one man gently by the leg until it was released.

Occasionally distress calls cause the rest of the school to flee from a wounded dolphin. Why this should happen is not known.

The Portuguese man-o'-war, known in places as bluebottle from the bluish tinge on its bladder-like float, trails long tentacles armed with stinging cells lethal to small fishes. Yet the young Horse mackerel seek safety among these tentacles at the first sign of danger.

are not detrimental to either species and may benefit one or both, as in the relationship of **phoresy**, in which one animal is carried by another (see p 193). The forms of these partnerships tend to run into one another and they are described by a number of terms which overlap in meaning.

Symbiosis means 'living together'. It is the most general term covering animal partnerships, although it has been used in the past in a more restricted way, to describe a partnership in which both species benefit. Nowadays, the terms **mutualism** and **commensalism** are used to describe partnerships of mutual benefit, or at least a partnership in which no one side suffers great disadvantage. In mutualism, there is an intimate relationship, in which the two animals are dependent on each other through an exchange of metabolic products. *Convoluta roscoffensis*, described in Chapter 3, and its algae are mutualists. Algae are also present in the tissues of coral polyps, where they assist in the formation of the chalky skeleton of the coral. The skeleton is made from calcium and bicarbonate ions removed from the sea water and combined in the coral's tissues to make hard, insoluble calcium carbonate. Carbon dioxide and water are liberated in the process. The algae use the carbon dioxide in photosynthesis and, by draining off the carbon dioxide, speed the uptake of calcium and bicarbonate by the coral's tissues. Coral grow more slowly if kept in darkness where the algae cannot photosynthesize.

Commensalism means 'eating at the same table'. Remoras or clingfishes, which attach themselves to the bodies of sharks, share scraps of the shark's food. A well-known example of commensalism is that between the Hermit crab *Pagurus bernhardius* and the Sea anemone *Calliactis parasitica*. The Sea anemone lives on the crab's shell and takes scraps of its food. It also protects the crab from enemies. They are sometimes joined by a ragworm, *Nereis fucata*, which lives inside the crab's shell. Another Sea anemone-Hermit crab association has reached mutualism: *Pagurus prideauxi* relies on a Sea anemone *Adamsia palliata*, for protection — the Sea anemone forms a cloak over the crab's body and neither can survive alone.

Commensalism can include partnerships based on shelter and protection. Small birds nest under the nests of birds of prey or wasps to gain protection, and many animals are commensal with man — as is suggested by the names of House sparrows, House martins, House mice and House spiders.

Inquilines are commensal animals living with social insects. A great variety of animals live in ants' nests, taking food but doing little significant damage. They include spiders, woodlice, beetles, crickets and millipedes. Some look remarkably like the ants they live with, so much so that sometimes they successfully beg food from the ant workers.

Chapter Eleven
Animal Languages

information and signals — visual signals: display: facial expression: gesture: posture: movement: illuminated signs — sound signals: function of: call notes: individual recognition — birdsong: and breeding cycle: duetting — drumming: whistling — mammal sounds — ultrasonics — musical insects: stridulation — bees' dance — croaking frogs and toads — sounds of fish and whales — chemical signals: pheromones: insect chemical signals — touch — electromagnetic language — high-level communication

Animals are constantly communicating with each other. By the way they move, by their postures, their gestures, the sounds they make and by their odours they are telling each other about the location of food, danger, territorial boundaries, or their need for a mate. This information is passed not only between members of a family but between larger social groups. Language, whatever form it takes, is the primary link between individuals, and the information passing between them binds them into an organized unit, just as nerve impulses and hormones co-ordinate the parts of an organism. The amount of communication that passes between animals varies enormously from species to species. Perhaps only those that reproduce asexually have no need for communication, but in all others there must be at least a minimum of communication between the sexes for breeding to take place. The need for communication increases as social organization becomes more complex until at the top of the scale, man communicates abstract concepts by means of a very sophisticated language.

Language is the systematic means of communicating ideas or feelings by the use of conventional signs, gestures, sounds and so on. For us, language has come to mean the verbal

For us, language has come to mean the verbal language of speech, but animal communication takes many forms, including the use of chemical signals (p 254-7). Most mammals mark their territories with scent marks using special glands. In some of the hoofed mammals conspicuous facial glands can be seen. Here a blackbuck is performing the delicate task of inserting a twig in one of its facial glands in order to leave its distinctive scent for demarcating its territory.

language, or speech, that has contributed so much to our advancement because it cuts out the necessity of each individual learning everything afresh; he or she receives knowledge acquired by others. For animals, language is more concerned with the communication of feelings, and there is an obvious comparison to be made between the language of animals and the primitive language of our emotions. The point was made at the end of Chapter 2 that, in our emotions, we are behaving very much like animals. Not only are the emotions the same for both, the expression of these emotions is similar. The smile of pleasure and the snarl of rage have their counterparts in animal language. When we grin, as opposed to a smile, the showing of teeth denotes a degree of anxiety or embarrassment. Monkeys and chimpanzees grin when anxious and wanting to appease a dominant congener. Furthermore, the 'sign language' of our emotions is instinctive, and we react instinctively to these emotions. It takes self control to hide expressions of rage and to avoid reacting to rage by becoming enraged or frightened, as appropriate, in return.

The basic information that animals have to communicate to each other can be listed quite simply. Of greatest importance for breeding is the identification of the species. Thus, courtship signals are usually species-specific, that is, the animal transmits signals peculiar to its species and only animals of its own species react. Secondly, the animal must announce its sex. It is a waste of time for a male to court another male. Moreover, the physiological state must also be communicated. A male bird does not sing and so tell other birds that he is ready to mate unless he is in breeding condition. The exact location of an animal is another piece of information which must be passed, if it is not already obvious. This is particularly important among territory holders. In social animals, notification of rank is another important message. Then there are such items, of lesser importance to some animals than to others, as alarm signals or warnings to predators that an animal is dangerous.

Much of the language, therefore, is linked directly with their physiological condition. The male bird sings only when his reproductive organs are ripe and are secreting sex hormones, and the castrated cockerel or capon lacks the rooster's distinguishing red wattles because the growth of these is dependent on the presence of the male hormone. An aggressive mammal is made ready for action through the secretion of adrenalin into its bloodstream; one reaction is that its hair rises. Raised hair is therefore a sign that an animal is likely to attack and raised 'hackles' have become part of the language of dogs.

The means by which signals are received include nearly every kind of sense organ in the animal kingdom. This is something of which we are only now becoming aware. The acute sense of smell of some animals is largely beyond our comprehension, and the electric sensibility of some fishes was quite outside our experience until the invention of suitable detection equipment, so the bulk of our knowledge about animal languages and communication concerns their use of visual and auditory signals. In this chapter, analysis of these signals begins with the visual signals of Herring gulls and continues with discussion of the languages of other animals. However, we should bear in mind that examples of animal communication can be found throughout the book.

Visual signals

The study of animal languages consists, in essence, of recording the signs that animals make in specific situations and, from a knowledge of the animal's motivation (fear, fight, or sexual drive, etc.) at a given time, deducing the meanings of these signs. Much of the pioneering work of analysing displays (that is, visual language) was carried out by Niko Tinbergen and his students on the Herring gull, which is why there is such frequent reference to this particular bird here. They found that, at the beginning of the breeding season, male gulls become increasingly aggressive. Each stakes a claim to an area of ground within the colony. Here, he will court a female and raise a family and, to do so in peace, he keeps the borders closed to all intruders. Despite his aggressiveness, however, real fights rarely break out. Encounters with intruders, particularly with owners of adjoining territories, are settled with threat displays in which the protagonists make their intentions clear by a simple code of postures without the need of wasting energy on actual fighting.

The Herring gull's threat displays are made up of intention movements of attack and flight which have become ritualized into signals, together with some displacement activities (see p 56, Chapter 2). The origin of the actions lies in the motivation of the gull as it approaches an intruder near the boundary of its territory. Let us think of this anthropomorphically: we are going down the path towards a tough-looking stranger at the gate. This leads to a conflict of emotions: aggressiveness towards a possible trespasser, and fear of what that stranger may do. The same conflict of emotions is at work in the Herring gull's brain, and the resultant of the conflict depends on the situation. If the bird is sure of himself, aggressiveness predominates, so a gull with an established territory has an overall aggressive motivation towards a casual intruder (see p 109). But once he has chased the intruder over the boundary he will begin to lose the confidence that his territory gave him. Fear will then predominate and he will retreat. The combination of aggressiveness and fear is known to ethologists as **agonism,** and the ensuing patterns of behaviour are called **agonistic displays.**

Visual displays are very common in birds, and they are emphasized by the development of special plumes and coloured patches of feathers that exaggerate or draw attention to body movements. The scarlet epaulettes of the Red-winged blackbird (see p 52, Chapter 2) are one example of this. The superb plumage of the Birds of paradise has long been prized by New Guinea tribesmen, as well as by Europeans (Magellan brought some back to the King of Spain in the fifteenth century). The elaborate fluffy plumes, long 'wires', epaulettes and gaudy colours seem almost to be the result of evolution 'gone wrong', because it seems that such elaborate plumage goes beyond what is needed for the survival of the species. The bizarre plumage is found in polygamous species which display on communal courting grounds, the male taking no part in rearing the young. It has been suggested that the fantastic plumage acts as a very species-specific signal to prevent interbreeding among very closely-related species. Yet fourteen hybrids have been recorded among the forty species in the family. By contrast, there are some Birds of paradise with quite plain plumage. These are monogamous and the males help to rear the young. So although no satisfactory explanation has yet been advanced, it looks as if the extraordinary plumages worn by most Birds of paradise may be connected with their polygamous way of life.

In all birds an unmated male responds to an approaching female as if she were an intruding male, and attacks her. She has to make her identity known and avert the male's aggression by adopting appeasement displays. A male gannet sits on his nest and advertises for females by bowing with wings held out, but when a female lands by him he grabs her neck with his bill. She may attack or retreat, but if she is

Territorial aggressive display of the Herring gull showing intention to fight: the beak is down-pointed ready to peck and the wings are lifted slightly in readiness to fly forward.

interested in him, she turns her head and deliberately presents her nape in a gesture of submission. Eventually, the male relinquishes the nest to her. Thereafter, whenever the female comes to the nest she gets her neck bitten, although this is followed by a greeting display in which the two gannets sit breast to breast and fence with their bills.

Visual signals are found in all animals with eyes, as in the Fiddler crabs (see p 140), fireflies, lizards and in many fishes. In mammals they may be controlled by the same conflicting emotions as have been described for Herring gulls. Konrad Lorenz has described the facial expressions of dogs adopted at different levels of aggression and flight motivation. Aggression is shown by baring the teeth and wrinkling the nose. The opposite flight tendency is indicated by laying back the ears, narrowing the eyes and pulling the corners of the mouth back. In a situation of both high attack and flight motivation, such as occurs when a dog is cornered, the well known fierce expression is presented: teeth bared, snout wrinkled and ears flattened. The same is seen in 'typical' pictures of Wild cats. Similar expressions are seen in horses and domestic cats, except that the laying back of the ears is a sign of aggression in the horse. Cats tend to squat when on the defensive, but when both attack and defence motivations are very high, we see the typical posture of a cat threatened by a dog: ears flat, teeth bared, back arched and tail raised in a 'bottlebrush'. This is very different to what we see in dogs.

An important difference between cats and dogs is that the former are solitary and the latter social. Whereas a cat attempts to defend itself and flees if beaten, a dog shows submission as part of its defence. Defence implies that an attempt will be made to repel an attack. Submission implies there will be no attempt at defence. The dog has to continue to live in the same social group as its attacker; to turn away aggression it shows its submission — by rolling onto its back. This is complete surrender, as it exposes its vulnerable parts and is at the same time powerless to use its own weapons. The effect of the gesture is to inhibit the victor's aggressiveness. Parallel displays in human warfare are seen in the raising of the hands or the surrendering of a sword. A supreme gesture of being at the enemy's mercy was the wearing of halters by the six burgesses of Calais when they surrendered to Edward III.

Signalling a threat by showing off the canine teeth is performed by other mammals, such as baboons and the

Pronghorn's heliograph
The pronghorn is sometimes called the Pronghorn antelope, but although antelope-like it is in a family all of its own among the hoofed animals. One peculiarity is that its horns are hollow and shed annually. Another is a deep sense of curiosity which hunters exploited.

A hunter would tie a white handkerchief to a stick, push the stick into the ground and wait for a pronghorn to come within gunshot to investigate. This sense of curiosity has led to the four million pronghorns the early French-Canadians discovered on the sage-dotted prairies of the Northwest being reduced to a few hundred thousand today.

This great reduction in numbers is the sad sequel to a neat adaptation for survival that was successful until modern man with his firearms arrived on the scene. The pronghorn is the speediest land animal on the American continent, rivalling the cheetah of the Old World for speed, in some ways outstripping it. It travels at an average of 30-40 mph but can step this up to 50, 60, even 70 mph over short distances. So it could afford to allow its curiosity to lead it to investigate anything strange because it could make a speedy getaway from natural predators. Moreover, as it fled it would use a natural signalling system to flash a warning of danger to all other pronghorns in the vicinity. As the pronghorn flees, covering the ground in leaps of 12-20 ft, its white rump patch flashes like a heliograph, catching the eyes of other members of the herd grazing in the area.

In keeping with its other man-like qualities, the chimpanzee signals its moods and intentions with its facial expressions. These five studies show a chimpanzee's face (top and left to right) in repose and pensive, showing excitement such as fear, and when happy as in play; (bottom) the surly appearance associated with a threat display and the greeting pout.

hippopotamus. In the hippopotamus, what appears to be an indolent yawn is really a danger signal. Tail wagging has a similar signalling function. Many rodents, including Ship rats and gerbils, quiver their tails when generally excited. The male Green acouchi, a large South American rodent, raises his tail, which is tipped with a tuft of white hair, and waves it, during courtship. By contrast porcupines lash their tails in threat. This lashing is made the more impressive by the specialized hollow-tipped spines, on the tail, which clash together. There is a similar contrast between tail wagging in cats and in dogs. All tail wagging may have started as a result of general excitement, as when a cat is stalking a mouse, indicating a general tension in a hostile encounter. The social dogs, however, have ritualized tail wagging into a friendly gesture, while the solitary cats, whose meetings with their fellows are more likely to be aggressive, use it as a threat.

Illuminated signs
The fireflies and the glowworm are neither flies nor worms. Both are beetles and, of over a thousand species of the family, the glowworm is the one British species. It is called a worm because the female is wingless and rather worm-like.

245

Fireflies proper on the other hand, are winged in both sexes. Female fireflies use their lights to attract males, emitting either a steady or a flashing light. The male North American *Photinus pyralis* flashes every six seconds and the female replies by flashing exactly two seconds later, so identifying herself to him. In North America this species has been given the name of Lightning bugs. The South American *Phrixothrix* is called the Railway worm because it has two rows of green lights down its body and a red 'headlight'. Firefly light organs consist of a light-producing chemical, luciferin, stored under a 'lens' of transparent cuticle, with a reflector behind it. An enzyme, luciferase, helps to oxidize the luciferin, to produce light. This is very efficient because 95% of the energy is converted into light, as compared to a light bulb in which 90% of the energy is wasted as heat. A firefly's light is equivalent to only 1/40 candlepower, but it appears brilliant and can be used for reading because the human eye is particularly sensitive to its wavelength.

A large number of deep-sea fishes use the same chemical reaction to produce light. The luciferin is contained in what are called photophores, or light-bearers, each of which is backed by a reflector and has a lens, sometimes with a diaphragm for altering the amount of light emitted. About two-thirds of the fishes in the twilight zone, about two thousand feet deep, where sunlight entering the sea peters out, possess photophores, often studded over the body in lines like the portholes on a liner. Very little is known about the habits of deep-sea fishes, but it is presumed that the patterns of light are used for identification and, perhaps, to give an indication of readiness to breed. Some of these lights may be bright enough to light up the sea around, for finding prey, while viperfishes and some deep-sea anglerfishes have luminous lures for enticing prey to swim near their mouths.

Deep-sea cod and rat-tails of the abyssal sea bed have lights powered by luminous bacteria. Krill, the surface-living shrimp-like creatures that form the food of giant whales, have several photophores, which are probably used as signals to help keep the vast swarms together.

Sound signals

Many animals make use of sounds to communicate. Birds use alarm calls to give warning of an approaching predator, and lions roar in threat. The most common uses of sound are to advertise the position of an animal, so as to attract a mate or repel a rival, and also to keep groups of animals together.

Tree shrews whistle in turn

Some birds sing antiphonally, that is one sings solo and then stops as another one takes up the song. A few years ago some American scientists studied Tree shrews on the Philippine Island of Palawan and found that they whistle antiphonally. Apparently they do this if disturbed or if they sense danger. They also do so just before retiring at night. The whistle is high-pitched and long drawn out and like so many of the alarm calls of so many birds it is difficult to locate. There would be an added advantage here that two Tree shrews in different places whistling antiphonally would confuse a predator, which would soon not know which way to look for its intended victim.

Varying the tune

The California ground squirrel gives a low growl expressing defiance, a sharp squeal indicative of pain or fear and a high-pitched squeal ending in a trill when being pursued by others of its kind. Apart from these and a few similar vocalizations that signal other circumstances the Ground squirrel conveys its messages by using a chirp that is varied in pitch, loudness and inflection.

When a bird of prey comes into view, the first squirrel to see it makes the chirp short and very loud, then bolts to cover. Other Ground squirrels hearing this repeat the call and also seek cover. For a ground predator such as a coyote or dog the chirp becomes trisyllabic. Other squirrels hearing this are warned, not to make a blind rush for cover but to take stock, seek to locate the enemy and act tactically to suit the circumstances.

Snakes are as much to be feared as a coyote, perhaps

more so, and a Ground squirrel seeing a snake edges up to within a foot or two of it. Having made certain it is not merely a dead stick it chirps in a low vibrating way, starting with a loud staccato note and following this with a series of subdued notes. At the same time it flags its tail. The chirp alerts neighbouring squirrels. The tail shows more or less where the snake is positioned.

Sounds are also used for identifying individuals, particularly parents and young. Sound signals are useful because they can go round corners, and the sending and receiving animals need not be in sight of each other. Call notes are particularly useful for keeping in touch in dense vegetation, whether by small birds in the foliage or shrews scampering among fallen leaves and grass stems.

Members of the same species may sound very much like one another to us, just as to us they look alike, but they can recognize each other by sound as easily as we recognize differences in human voices. Proof of this is not easy to come by, but recordings of the calls of terns, gulls, gannets and penguins show that the calls of each individual have characteristics that enable them to be distinguished from the calls of other individuals. Moreover, it has been shown that tern chicks react only to the voices of their own parents and Niko Tinbergen records that a sleeping Herring gull awoke when it heard the calls of its returning mate, whereas it showed no response to the calls of other gulls which were indistinguishable from those of the mate to the human ear.

247

Bird song

Undoubtedly, the most familiar kind of acoustic communication among animals is bird song. Apart from the sheer volume of noise that can be heard early on a spring morning, the pattern of notes is aesthetically pleasing. Whether there is any significance in bird song appealing to our sensibilities is a matter of conjecture, as is the possibility that birds themselves actually enjoy singing. Not all birds sing: vultures, for instance, are silent and many other birds utter sounds that are not musical. Technically these may be songs in that they are vocal advertisements; they are referred to by ornithologists as 'songs' even if they are no more than a harsh croak. As a general rule, the loudest songs come from birds with inconspicuous plumage, or those that live in dense vegetation. Songs are best developed in the suborder Oscines of the perching birds, order Passeriformes. These birds are popularly called 'song birds'. They include the familiar birds of garden and countryside as well as bowerbirds and Birds of paradise, and make up nearly half the total number of the nine thousand bird species. The Oscines have five to eight pairs of muscles in the voice box or syrinx, the bird equivalent of our larynx. It contains vocal cords actuated by the small muscles. Other perching birds have four pairs or less. Birds do not modify sounds produced in the voice box with the tongue and mouth as in human speech. The variety of sounds is made possible by the intricacy of the syrinx. Some birds sing more than one song at once. Reed warblers and Brown thrashers sing two songs at once; the American Wood thrush sings four notes at once.

A form of song is the drumming of woodpeckers. At one time it was thought that the Great spotted woodpecker of Europe and others made their drumming sounds vocally, but films have shown that they actually drum on the wood with their bills. A different form of instrumental music is the whistling of American woodcocks, which dive and soar during courtship. The slipstream plays through special stiffened flight feathers which whistle as they vibrate. The drumming of the European snipe is produced similarly by air rushing past two modified tail feathers. Both of these appear to be substitutes for vocal song in the birds' courtship.

As with other activities concerned with reproduction, singing is linked with the physiological changes in the bird's body. As the testes swell and start to pump out sex hormones in the spring, singing gets under way. However, there are a few birds, such as the mockingbird and the European wren,

The male snipe is said to sing with its tail. Its courtship includes aerobatics within sight of the female. A characteristic performance is for the snipe to go into a near-vertical dive with one feather on each side of its tail extended and vibrating in the wind to produce a drumming or bleating sound. Sometimes the female drums also. The drumming may be heard by night as well as by day.

Telling about food

Most animals are markedly egocentric in the matter of food. One that finds food tends to 'hog' it, keeping it to itself, even fighting off others that try to share it. It could hardly be otherwise, since adequate food is the first step to survival. Yet it is not unusual to hear people talking of the birds they feed 'telling the others'. They base this on having seen a bird take some food, fly off with it, and when that bird returns for more others come flying in. To some extent the process is visual. Birds are very quick to see when one of their fellows has food and to watch where it is getting it.

Some species do 'tell'. When one puts out food for

birds and there is a starling on the roof, that one calls on a rising note and within seconds other starlings that had been previously out of sight come flying in to share it.

Some gulls go further and announce what kind of food. There is a simple test for this. Throw bread to gulls and those near enough to see it come flocking in. Throw bread with some meat or fish among it and gulls will come flying in from a quarter of a mile radius and instantly, because the first gulls to find the meat or fish make a special call.

We see this special call functioning when a shoal of small fishes swim near the surface, in the sea or in a river. The instinctive special call is uttered by those in the vicinity and in no time gulls come flying in from all quarters to prey on the fish.

which sing all the year round and a few females, the mockingbird and the European robin for instance, sing as well as the males. There may even be a duet between male and female. The Eastern whipbird of Australia is so-called because the song starts with the male uttering a note that increases in intensity and ends with a sharp whipcrack. The female's reply is of two or three notes. The function of the duet is to keep the pair informed of each other's position while feeding in the dense undergrowth. A familiar duettist in East Africa is the bellbird or bell-shrike. At first it appears that only one bird is singing, so well timed are the two songs and, if one member of the pair is absent, the other will complete the whole song sequence itself.

The main function of bird song is advertisement. Each male sings from a number of positions to show the extent of its territory. This also keeps other males at bay while unmated females are attracted by it for courtship. Species-identification is part of the advertisement and the physically similar Eastern and Western meadowlarks, whose ranges overlap in North America, are prevented from interbreeding by their distinctive songs. The Western meadowlark has a flute-like song, the Eastern species has a slurred whistle.

At the start of the breeding season the male's song also has an important function in helping to bring the female into reproductive condition. Singing generally wanes after the eggs have been laid, but continues to some extent as a means of maintaining both the bond between male and female and the integrity of the territory.

Musical insects

A large number of insects make noises as a by-product of the rapid beating of the wings in flight; these noises are usually quite incidental and probably functionless. The mosquitoes, however, have taken to using the wingbeat buzzing of the females as a beacon for attracting males (see p 129, Chapter 6). Elsewhere in the insect world there is a general sensitivity to airborne sound through hairs on the body, which vibrate sympathetically with the sound waves. In some species, it is not easy to decide whether the insect is responding to airborne sound or to vibrations through a solid medium. The Death watch beetle lives in old timbers in houses. It has long been the subject of a superstition based on the tapping the adult makes with its head on the woodwork. Heard at night, it sounds like the ticking of a watch and was held to foretell death in the household. In fact, the sound is made by both sexes and appears to be a mating call. But whether the insects pick up vibrations through the wood or actually hear the sound is still a matter of doubt.

In only a relatively few insects are there well developed hearing organs. The ears of insects are more properly called 'tympanal organs' because they are not very much like human ears, either in structure or functioning. Tympanal organs are found in moths, where they are used to detect the ultrasonics of hunting bats (see p 64, Chapter 3), and in cicadas, a family of bugs noted for their shrill, piercing calls, as well as in the crickets and grasshoppers, all of the order Orthoptera. Strictly speaking, the jumping orthopterans are classed as Short-horned grasshoppers (Acrididae) with short, stout antennae, and Long-horned grasshoppers or Bush crickets (Tettigoniidae) with very long, thread-like antennae and crickets (Gryllidae) with filamentous antennae and long cerci at the rear of the abdomen. In each of these groups, sound-producing organs are present in the males and used for attracting females, while the sound-receiving organs are possessed by both sexes.

Early in this century, the entomologist Regen showed that female crickets took no notice of a male hidden under a glass

Ultrasonic hamster
Most of us are prepared to believe now that the sun does not revolve round the earth. We are, however, still inclined to base our thinking on the notion that the living world revolves around us. When, therefore, it was discovered 30 years ago that bats make sounds too high for us, the lords of creation, to hear we were moderately surprised. And to think they were using these ultrasonics for echolocation in the same way as our latest inventions were doing was quite astonishing.

The rest of the story has crept on us so slowly it is only when we stop and think that we realise that the discovery about bats was only a beginning. Indeed, a new world was about to be opened up. Soon we were learning that porpoises and dolphins were using ultrasonics to locate food, find their way about and communicate with each other.

The field widened even more when shrews were found to be using calls higher than the ears of most of us can detect. Then it was the turn of the rodents. Baby mice were caught in the act of 'talking' to mother in ultrasonics, which probably even the acute ears of cats cannot detect. After this it was learned that adult rats and mice of many species utter menaces just before a fight, in sounds above our hearing. Some people, especially young people, can just hear these. There are people with exceptional hearing that can hear the high-frequency calls of a pet hamster sitting on their shoulder; but although millions of people have kept hamsters during the last 40 years, this fact has only recently been recognized scientifically.

Rodents grit their teeth

No rabbit, hare or rodent need ever have dentures. They have teeth that are always growing at the roots, at an alarming rate when viewed as cold statistics of tooth-growth. This is especially true of those characteristic gnawing teeth, the incisors so prominent at the front of the mouth.

In the North American porcupine each of the incisors may grow $3\frac{1}{2}$ inches in length a year. The Guinea pig tops this with 10 inches, the European rabbit with 5 inches and the Common rat with 4-5 inches a year. Even these breathtaking figures are put to shame by the Pocket gopher whose front teeth increase by an amazing 9-15 inches a year, and a fullgrown Pocket gopher is only $14\frac{1}{2}$ inches long including its $4\frac{1}{2}$ inch tail.

Skulls of these animals are sometimes found portraying the actual cause of death. In these, one or more of the incisors has become displaced so that instead of being worn down by constant chewing, the teeth have grown unrestricted and have curved up and round to penetrate the skull, locking the jaw so that starvation results.

All rodents are hearty feeders, perpetual chewers. Squirrels chew metal nameplates on trees, rats chew lead pipes, even concrete. In addition, it has recently been observed that rats are always gritting their teeth when they have nothing else to do. This ensures constant wear as well as keeping a sharp chisel edge to the incisors. It probably also eases boredom, like the patience beads of the Arabs. It may even serve for communication, like the supposed tummyrumblings of elephants, now known to be a form of communication. Who knows? The subject has not yet been investigated.

bowl where they could see him but not hear him. When a microphone relayed the male's chirruping, the females gathered around the loudspeaker. They locate the source of the sound by means of hearing organs which are situated on the tibia of each foreleg in both crickets and Long-horned grasshoppers. Short-horned grasshoppers and cicadas have hearing organs on the sides of the abdomen close to the thorax.

Like the so-called ears, the voices of grasshoppers, crickets and cicadas are not the same as the vertebrate counterpart. They are more in the nature of instrumental music. Cicadas are the loudest of all insects and can sometimes be heard for a quarter of a mile. Close to, they may make reasonable conversation impossible. The noise-making mechanism consists of two tymbals, membranes on each side of the abdomen which are pulled out of place by special muscles and allowed to snap back with a click, rather like the lid of a biscuit tin being popped in and out and allowed to vibrate freely. At a rate of one hundred to five hundred times per second, this produces the deafening buzz which can be changed in tone and volume by secondary muscles altering the tymbal membranes.

Crickets and grasshoppers are said to **stridulate,** producing their sounds by rubbing hard parts of the body together. Short-horned grasshoppers rub the knobbed inner surface of the femur of the long hind leg against a vein on the narrow, leathery forewing so that the wings vibrate at a natural frequency. Crickets and Bush crickets rub the two forewings together. Temperature affects the chirping of grasshoppers and crickets and such is the strict relation between temperature and the rate of metabolism of coldblooded animals that the air temperature in Fahrenheit can be calculated by subtracting forty from the chirps per minute, dividing by four and adding fifty. An easier but rougher method is to add forty to the chirps counted in fourteen seconds.

Each species of these insects has a 'signature tune' by which it can be identified as surely as an ornithologist recognizes a bird by its song. Only virgin females respond to the calls; fertilized females do not respond and each species may have a variety of songs, some having as many as thirteen. The normal song of the male cricket attracts females and also attracts other males so that they gather in a group where the females can find them more easily. When a female arrives nearby, the male sings a courtship song; he has other tunes to play to quieten the female if she becomes restless during mating or if another male disturbs them.

The honeybees' dance

One of the most extraordinary and complex systems of animal communication is the 'waggle dance' of honeybees. Some thirty years ago Karl von Frisch discovered that bees can communicate the position of a source of pollen and nectar, such as a bed of flowers, to their fellow workers when they return to the hive. They do this by waggle dancing on the honeycomb. The exact nature of the messages has not been fully worked out and there is even some dispute as to the main mechanism employed, because the senses of touch, smell and hearing are all involved in the passing of information between bees.

When a honeybee worker returns from a newly-discovered source of food within fifty yards of the hive, she performs the 'round dance' on the comb, circling among other workers who jostle her, so informing themselves which dance she is performing. They also pick up scent from her body. The scent of flowers clinging to the bee's body or passed around during food sharing are the only clues given about the actual food source. Since it is, in this instance, near to the hive they can go out and look for it. For sources farther from the hive, a 'waggle dance' is used. This makes a figure-of-eight. The orientation of the dance is important as it shows other bees the direction in which to fly. If the bee moves vertically up the face of the comb, the food is in the direction of the sun. If it runs vertically down, the other bees must fly away from the sun. Similarly, if the dance is orientated at thirty degrees left of vertical, then the bees must fly a course set at thirty degrees left of the sun's position.

As the dancing bee goes down the centre of the figure-of-eight, it 'waggles' its abdomen, the number of 'waggles' being inversely proportional to the distance between hive and food. Presumably the other bees detect the frequency of the 'waggles' by jostling the dancer. They also detect a low buzzing, a series of pulses the frequency of which is also related to the distance the foragers must travel to find the food. The amount of dancing depends on the quality of the food source. A bee that has found a plentiful supply of nectar dances vigorously and recruits many others to help gather it but the discovery of a poor source may not lead to any dancing on the bee's return.

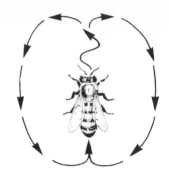

Diagrammatic representation of a returning forager honeybee performing a round dance (above) signalling nectar less than 50 m from the hive, and the waggle dance (below) indicating the direction of a source of nectar at a greater distance.

Purring elephants

It never does to jump to conclusions. In the early part of this century elephants had big game hunters puzzled. When shooting elephants with cameras replaced shooting them with guns, the naturalists were also puzzled and like the big game hunters came home with stories about the tummy rumblings of elephants.

It was not the rumblings as such that puzzled them, because with such an enormous beast with a big intake of food daily they were to be expected. The puzzling thing was that the elephants seemed to be able to control them and the tummy rumblings of a whole herd would stop as if at a signal.

The truth has been discovered only a few years ago. The sounds actually come from the elephant's vocal cords. An elephant finding itself hidden from its fellows while feeding starts this rumbling sound in its throat. The sound reverberates throughout its massive body. It is the method elephants use for keeping in touch with each other. So long as one elephant can hear another 'purring' it knows it is not alone. Moreover, should an elephant sense danger of any kind it stops purring. Other elephants around know this is a danger signal and they stop also, so the alarm spreads throughout the herd.

So long as elephants are in the open and each can see what the others are doing, they have no need of this bush telegraph. Only when they are among thick cover are the rumblings used.

The male Tree frog advertises its presence to the female by a loud croaking, blowing out its throat-sac like a balloon.

Croaking frogs and toads

The chorus of insect tunes is matched in strength, persistence and often monotony only by the croaking of frogs and toads. Throughout the breeding period, the males of these amphibians gather in ponds and streams and croak unremittingly. A typical example is the bullfrog, native to the eastern United States and introduced to the western States as well as to Hawaii and Cuba; it makes a call that has been described as sounding like a man with a deep voice shouting 'rum' into an empty barrel. The sound is repeated in groups of three or four every five minutes and is made, as in all frogs and toads, by air being passed to and fro from mouth to lungs across the vocal cords. Some of the air enters airsacs in the floor of the mouth which inflate and act as a soundbox or resonator. The croaking is used to attract females or repel other males, but there are several calls, serving a variety of purposes. The bullfrog has a mating call, which also helps to bring other bullfrogs to one place, a territorial call, a release call made by either a male or an unreceptive female when clasped, a warning call when danger threatens, and a distress call when hurt. In many other species of frogs the calls are used in much the same way as this but in one treefrog, at least, it seems there is a sort of 'lek' situation, such as is found in birds. The females are attracted to one particular male in a group, who croaks longer, louder and more rapidly than the rest of the chorus.

The sounds of the deep

It was well known to the old-time whalers that whales not only had very sensitive hearing but that they made noises. When 'going on' a whale to harpoon it, the greatest care had to be taken to avoid knocking anything in the boat and so 'gallying' or scaring the whale. On the other hand, the wooden hull of a whaling boat, such as was then in use, made an excellent transducer, converting the waterborne sounds into airborne sounds. The whalers could hear the whistles, creaks and trills of the whales under the boats, especially of the White whale, which became known to the whalers as the Sea canary, because it was so vociferous. Nowadays, hydrophones and recordings have revealed the 'songs' of the whales for all to hear. The so-called 'song' of the Humpback whale is musical, but it is not a song in the same sense as a bird's song in that it does not advertise a territory or repel other males. As far as is known, whales' songs are beacons that convey the position and identity of the one that makes them. Due to the good conduction of sound in water, they can be heard for hundreds of miles, although the range is now reduced by interference from the continuous beating of ships' propellers.

The hydrophones that revealed so much of the whale sounds have also shown how frequently fishes produce sounds. This became an important study during World War II when it was necessary to distinguish between the sounds made by fishes and those made by enemy submarines. Fishes produce noises either by stridulation or by using the swimbladder as a resonator, or by a combination of the two. Rasping noises are produced by grinding the teeth and Sea horses click by nodding their heads, when a projection on the skull catches on a bony plate. Stridulations are usually high pitched rasps, scratches and squeaks whereas swimbladder noises are low, 'hollow' sounds. In some fishes the swimbladder, which is a buoyancy organ, connects by a narrow tube to the intestine and noises are made by the passage of air bubbles, as in the American freshwater eel. In the John Dory, an oval, flattened fish of the Atlantic, sounds are made by muscles vibrating the swimbladder, but in other fishes the swimbladder is made to resonate by vibrations set up by the teeth or certain bones grating together, or by the fins knocking against the body.

In so far as is known, fishes use sounds to attract mates, warn of enemy attack or keep shoals together. Pinfish, however, use sounds to advertise their territories. These are small fishes with sharp spines on the back, living off the Atlantic coast of the United States.

Chemical signals

Because we have a poor sense of smell it is virtually impossible for us to tell what messages animals are passing by means of scents and odours. To some extent we are now helped by refinements in the technique of chemical analysis but we have, in the main, to deduce what is being communicated by observation and inference. We can see a dog mark a lamp-post and infer that other dogs find something interesting in the smell, but we have to guess that they are in fact identifying the owner of the smell.

Leaving scent marks about is a common behavioural trait among mammals. The mark may consist of excreta, urine, saliva or the productions of special glands, and marking may involve special rituals, like the leg-cocking of a dog or the tail-flapping of a hippopotamus, who spreads his dung over

The John Dory (see text) has a dark patch on the flank popularly referred to as 'St. Paul's thumb print', the legend being that this was the fish in which the apostle found the tribute money.

Singing fishes
Batticaloa is a small port at the head of a lagoon on the east coast of Sri Lanka. Visitors can go out with a local fisherman on a clear still night and listen to the singing fishes. Holding a stick to the bottom of the boat or in the water and placing the ear to the other end, they can hear sounds as of distant music, rising and falling, a twanging sound in differing tones.
Anyone hardy enough to dive overboard hears the 'music' as of an orchestra of strange instruments tuning up, playing softly but without harmony or melody, but without discordance, in a rhythmic throbbing.
This is only one of many sounds recorded for fishes. The sirens of Greek mythology are believed to have been based on the 'singing' of the meagre or weakfish. In modern times the knocking and purring of the Satinfin shiners have been fully investigated. The shiners are small, minnow-like fishes of North America, living in fresh water. Three types of sounds are made: single

254

knocks, series of rapid knocks and purrs. The knocks are used by males against rival males, especially against those straying into an occupied territory. The purring is associated with courtship and has a marked effect on the females, making them more receptive.

The sounds made by Satin-fin shiners, and probably those used by many other fishes, seem therefore to serve the same function as birdsong: to advertise possession of a territory, deter potential rivals and intruders and to enhance courtship.

Scent signals

Many mammals use a range of pheromones. The Blacktailed deer, a small deer of the Pacific coast of North America, has at least six pheromones which are used to convey messages.

The tarsal organ on the inside of the back leg consists of secretory glands and a tuft of hair, and its odours are used in mutual recognition. Throughout the day and night, members of a herd sniff each other's tarsal organs and strange deer are also smelt as a prelude to aggressive behaviour. Chemical analysis of the scent shows that it is different for each individual and that the scent can also be used to distinguish sex and age.

Scent from the metatarsal glands on the outside of the hindlegs smells of garlic and is secreted when the deer is frightened. When fleeing, deer leave scent from their interdigital glands in hoofprints. Forehead glands are rubbed against twigs and branches to mark the home range, particularly in the vicinity of the sleeping place. Urine is also used as a marker and in courtship and aggressive encounters. Finally, there are scent glands of unknown function on the tail.

the vegetation conveniently near the nose height of other hippos. Perhaps the origin of this behaviour is that the deposition of these substances loads the home area of an animal with its own smell. This makes it familiar, giving a sense of security, and enabling the animal to find it again if it strays. It is noticeable that mice quickly mark their cages after they have been cleaned, so that cleaning cages is a never-ending chore and overcleaning may lead to a dehydrated animal. The odour not only makes the mice feel 'at home' but other mice recognize that the territory is occupied and keep away. It also provides the mechanism for territorial defence; the owner derives self-assurance from it while the intruder feels out-of-place.

The addition of pheromones to these excretions and secretions increases the information they contain. Sex, receptive conditions of females, and rank may be conveyed quite simply and, because scents linger, scent marks are a convenient form of communication because the message can be left while the animal goes about its activities elsewhere.

Not surprisingly, scent marking is frequent in the breeding season and the cloven-hoofed deer and antelopes have developed a number of secretory glands and means of disseminating scent. Many species possess a pre-orbital gland just to the front of each eye. Scent is transferred to a firm object by rubbing the head against it or the head may be brushed against leafy vegetation. In a delicate operation that appears to endanger the eye, some antelopes wipe the orifice of the gland against the tip of a twig. The **thrashing** of saplings by deer, which causes so much damage to forestry plantations, is not done to clear the velvet from the newly-grown antlers, as is sometimes thought, but is a process of scent marking. To some extent, the amount of thrashing or **fraying** is dependent on the density of competing males. As boundary disputes increase when the number of males rises, thrashing becomes more common, but it may diminish at higher densities because the deer drive each other off by fighting rather than by warning signs.

With the exception of the social insects whose pheromone systems are described in Chapter 2, chemical signals in animals other than mammals are of a simple nature. They are effective because they are persistent, they travel beyond the range of vision and, perhaps most important, they require neither elaborate organs for their production nor well-developed sense organs for their reception. So, among the simplest forms of life, signals received through the

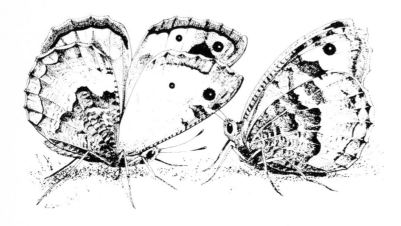

The male Grayling butterfly faces the female and clasps her antennae between his wings to bring them into contact with special scented scales, known as androconia (see text).

Thumper raises the alarm

The boy with his grandfather crossed the bridge and turned sharp right by the hedge into a field, surprising a buck rabbit, who raised his hindquarters and thumped the ground with his hindfeet. Grandfather, who in his time had done his share of poaching, explained that the rabbit was telling all the other rabbits somebody was coming.

Probably even grandfather would have hesitated to suggest that this thumper had deliberately acted this way to warn the other rabbits. Rather it was an automatic action the effect of which was to warn the others.

It has been suggested, improbable though it may sound, that when someone sitting with crossed knees waves the foot up and down impatiently, that person is showing a subconscious desire to walk away. Perhaps thumper is merely registering the intention movement to leap away.

Two scientists were studying gerbils, small desert rodents with long hindlegs, similar to jerboas. They were giving the gerbils small electric shocks applied to 'pleasure centres' in their brain. The scientists found that when they stopped giving them an electric shock the gerbils thumped with their hindfeet. They also noticed that in the free state gerbils stamped if interrupted when doing something pleasurable.

By the same analysis we could suppose the rabbit crouched in the sunshine, at peace with the world, when grandfather and grandson intruded.

chemical senses are used to bring the sexes together. Among insects, however, sophisticated chemical signalling systems have been developed. Some of the most notable users of scent are the moths. Although the signal is simple, being no more than a beacon for attracting males to females, the system is very sensitive. The feathery antennae of the male moth, contrasting so clearly with the small antennae of the female, act as a delicate net for trapping scent. The amount of bombycol, the scent of the female Silk moth, given out by one individual, is sufficient to fill a volume of air several miles long. On detecting just a few molecules, a male Silk moth turns and flies upwind, towards the female disseminating it. When he gets near her, the increasing concentration of scent guides him to her.

A more elaborate courtship by scent is shown by the Grayling butterfly of Europe. When ready to breed, the male takes up station on the ground and flies up as soon as a female flies overhead. If she is ready to mate, she lands and the male alights behind her. He then walks round to face her and, in a delicate, showy bowing ceremony, he clasps her antennae between his wings. Afterwards he walks behind her again and mates with her. The significance of the bowing ceremony is that the male brings the female's antennae in

256

contact with special scent-producing glands on his wings. The scent is a pheromone that brings the female into breeding condition.

Touch and tingle

The use of touch in communication has not been easy to analyse. Often it is difficult to distinguish touch from smell. When two ants meet and touch antennae, it is not easy to tell whether they are feeling or smelling each other. Touch appears to be important among some vertebrate animals in at least a generalized way. Snakes entwine their bodies during courtship and giraffes 'neck' by pressing their necks together. Mutual preening and grooming must also be stimulating, presumably in much the same way as we find caresses stimulating, and one could suppose that the head-on butting by male Bighorn sheep is a form of tactile communication.

The spiders are one group of animals in which touch is known to be important in both courtship and feeding. A spider's web is a glorified trip-wire, and as soon as the spider feels vibrations in the web it rushes out to seize the trapped insect. The male spider has to ensure that the much larger female does not regard him as food, so he identifies himself with a code of vibrations made by plucking the strands of the web. The female turns from thoughts of food to thoughts of love and the male presses home his advantage by caressing her with his legs.

One of the surprising discoveries of animal communication is that some fishes communicate by means of **electromagnetic fields** (see also Chapter 2, p 47). It has been known since early times that certain fishes can deliver an electric shock. The Electric rays deliver a discharge of two hundred volts and the Electric eel of the Amazon basin (not a true eel) can discharge five hundred and fifty volts, which is said to be sufficient to kill a horse, although it is more likely that it can stun a horse which then drowns. The Electric catfish of Africa is another powerful electric fish delivering three hundred and fifty volts and the marine stargazers produce fifty volts, although others such as the Elephant-snout fishes and the knifefish of Africa can deliver only weak currents.

There is nothing particularly revolutionary about fishes producing electric currents. Every time any muscle contracts there is a small electric discharge. In these fishes, certain muscles have become turned into batteries which no longer

257

contract but specialize in generating electricity. In the Electric eel there are six thousand electroplates, each of which generates one tenth of a volt.

The high voltages are used for killing or stunning prey or for defence, whereas the low currents are used for navigation and for social behaviour. The Electric eel has both high and low voltage discharges. The electric organs set up a field around the body of the fish like the field around a magnet. This field is sensed by special sense organs in the skin and the fish can detect distortions in the field caused by an object nearby. It is in fact a kind of radar by which electric fishes can find their way around at night or in murky water. The field of Elephant-snout fishes is maintained by a continuous series of discharges that is slow when the fish is resting but rises to a hundred per second when the fish is active. Others, such as the South American Gymnotid electric fish, have a continuous frequency discharge of lengthy pulses. Not much is known about the social lives of many of these fishes, but experiments suggest that they use their electric organs to communicate the same information as other signals, such as visual, acoustic and scent signals, convey. They are used for species recognition, sex recognition and for the maintenance and establishment of dominance hierarchies. So these fishes are living in an 'electric world' paralleled by our world of sight, sound and scent.

Of higher things

Communication between animals is usually prosaic, aimed at finding food, starting and rearing families and avoiding enemies. The information conveyed is factual and mechanical but there are rare occasions, among the higher animals, when they seem to be passing more than ritualized, species-specific signals concerning reproductive state or the position of a food source. They sometimes seem to express a thought or a concept. This is hard to prove but there are the odd instances like that of the female chimpanzee who approached her male companion, whimpering. He sat in front of her and pulled down the lower lid of one eye. After an inspection, he removed a speck of grit with a finger. How did the male chimpanzee know what the trouble was? There is no answer, as is the case with an experiment with chimpanzees in which one was shown some hidden food. It was taken back to its companions who were later released into the compound. They made for the hidden food without hesitation, apparently having obtained the information as to

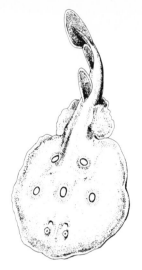

The Electric ray, or torpedo, one of the first fishes to be recognized as capable of dealing an electric shock. Its large pectoral muscles are modified to form electrogenic organs and produce a discharge of 200 volts.

Cat conveys a message
The owner of a Siamese cat was in the habit of taking his ease of an evening in a particular armchair. When he did so the cat, almost invariably, would leave the chair in which it was sleeping, go over, jump into his lap, and knead his jacket with its forepaws, while the man stroked the cat. So the cat would eventually settle, fully content to the point of continuing to purr while apparently fast asleep.
On one occasion the owner had missed using the chair for several evenings. Then he went into the room and stood pondering near the window. The cat left the chair in which it had been curled up fast asleep and tugged at his trouser leg with claws out, just two tugs before sitting on its haunches and looking up into his face with what can only be described as an appealing look.
Still somewhat absent-minded, the man then

walked over and lowered himself into his favourite chair. As he did so the cat jumped on to the arm of a neighbouring chair, leapt from there onto his lap and started the usual routine. Only then did the significance of what had occurred dawn on the man. The cat had so obviously decided it wanted him to sit in the chair so that it could enjoy the routine of kneading, being stroked and the rest.

A commonplace event to cat lovers, no doubt, but seldom could it be so clearly evident that a cat is capable of an abstract idea and of taking steps to put it into effect. And if a cat can do this with a human, there is no reason to suppose that cats, and other of the higher animals, cannot make their wishes known to their fellows.

its whereabouts without, so far as could be seen, any sound, gesture or other signal having been passed between them.

Somehow the chimpanzees must have communicated to each other, and in quite a sophisticated manner. That chimpanzees could conceivably have such high ability in communicating is suggested by Washoe, a female chimpanzee brought up by Allen and Beatrice Gardner as one of the family. They communicate with Washoe by means of ASL, the American Sign Language used by the deaf and dumb. Unlike the finger-spelling of British Deaf-and-Dumb language, this is a system of gestures, each of which stands for a whole word or concept. For example, 'flower' is described by holding the fingertips of one hand together and smelling them. Washoe now has a vocabulary of some one hundred and fifty signs, and invents her own signs and meanings. She learned to give the sign for 'open' when she wanted to go through any door and then, amazingly, used it when she wanted a water tap turned on. She also coined the phrase 'open-food-drink' for a refrigerator, and when she heard a dog bark she 'said': 'hear-dog'.

259

Chapter Twelve
Unusual and Intermittent Behaviour

window-tapping in birds — addiction — alcoholism — smoking —
drugtaking — weather-forecasting — earthquake prediction —
overkill — opportunism — cannibalism — sibling murder — pain —
akinesis — thanatosis — curiosity — charming — cats and catnip
— aberrant instinct — ant addiction — fossil behaviour — rain
dance — snake suicide — time sense — hedgehogs milking cows —
autophagy — hoarding packrats — birds' balance

So far we have dealt with the aspects of behaviour that form the basis of everyday living, and with those that change regularly with the seasons. In a few words, they concern those things we can expect to see. There are in addition features of behaviour that are intermittent or random. These are unexpected. They are things the layman notices most and which the professional scientist, on the whole, notes but tends to disregard, mainly because they cannot be systematically investigated.

We can draw a parallel here with human affairs. It has been said that history is not in the headlines. A daily newspaper or a radio or television programme seldom deals with the humdrum flood of events that carries humanity on to the unknown goal of history. Reflected in the headlines are the unusual and spectacular events, the nine-day wonders. By their very nature these events are isolated from the general rules of living; they tend to be inconclusive and are the subject of speculation.

Window-tapping

Take the case of birds tapping at windows. It is not every day that a bird taps on a window pane with its beak. Yet this is

Chimpanzees' rain dance
Heavy rain affects different beings in different ways. Pigeons, which do not bathe in the usual way, lean to one side and raise one wing vertically, rain-bathing. Sunbirds and hummingbirds get excited. Parrots go seemingly crazy, flapping their wings, screeching, even hanging upside-down to add to the confusion. Traditionally, the household Puss will rush up and down stairs and race round the house, although sheltered from the rain.

The most extraordinary display was seen by Jane Goodall when studying chimpanzees in the wild. She saw two groups of males rush up a slope. Then one charged downhill, followed by another, then another. They swung their arms and then hooted, sprang into trees, broke off branches and, on reaching low ground, turned, went up the slope and repeated the performance. They behaved in this beserk way for half an hour, while rain pelted down and lightning flashed and thunder growled. The females and young sat in the trees watching.

Jane Goodall called this the chimpanzees' **rain dance**, a piece of primate behaviour unknown until 1964, when she witnessed it. It seems to have something in common with certain small boys who, as has been known for a long time, seem to get highly excited when the rain teems down and long for nothing more than to go out into it.

George Schaller who studied gorillas in the wild noted that although they usually sheltered from heavy rains under trees, occasionally they were seen to go out deliberately into the rain or to build their nightly sleeping nests in the open where they received the full force of the pelting downpour. Although gorillas normally walk on all fours, and rarely go more than five feet bipedally, old males were seen to walk upright five times this distance in pouring rain.

something that happens, and every year hundreds, perhaps thousands of people experience it and comment on it. Sometimes a bird will tap persistently at the bright hubcap of a car, or at the driving mirror (see p 264). Then it may earn a secondary headline in a newspaper.

The usual explanation given by ornithologists is that the window, hubcap or driving mirror happens to be within the territory of a breeding pair of birds. The male sees his reflection and this, to him, looks like another male of his own species intruding on his territory. So he attacks, as he would in fighting a rival, with his beak. The truth of this explanation can be tested by placing a mirror upright on the ground within the territory of a breeding male. The male in occupation of the territory will approach the mirror, go into an aggressive display, and then attack.

So, to everyone who reports having a bird tapping at a window the scientist gives this stock reply and, satisfied, goes back to what he was doing when the questioner interrupted him.

'But', says the questioner, 'this one taps at the window every morning at 8 a.m. and I open the window and feed it. It is tapping for food.' The scientist sighs, looks up again from his research and says: 'The bird first tapped the window on seeing its image in the glass. You thought it was tapping to attract your attention, opened the window and gave it food. So it learned by this that if it tapped on the glass it would get a free meal.' Once again, satisfied that he has given the last word, the scientist goes back to his work.

There is another way of approaching this. If we collect the letters written to naturalists on this subject and analyse them, we find they can be arranged in three piles. The largest pile will be about birds that were, beyond question, attacking their own reflections, during the breeding season. The second largest will relate to instances that look as if the bird were deliberately begging for food. The third, the smallest, we will come to in a moment.

As we re-read the letters in the first two piles, we notice certain differences. In the first, when a bird is tapping belligerently, the observer reports that the attacking bird flies away as he opens the window and either goes off to attack another window or returns to the same window as soon as his back is turned. In the letters in the second pile we read that when the window is opened in response to the tapping the bird does not fly away. It may even hop through the window onto the sill inside. When it is given food, it eats,

263

flies away and may not return until the same time the next day. Also, the bird is as likely to be a female as a male.

How the bird first came to know it could get a free meal by tapping on the glass is an open question. What matters is that there is here a distinct possibility that this is a piece of insight behaviour (see p 31). Can we find supporting evidence that birds tap with their beaks, under unusual and unnatural circumstances, in the endeavour to satisfy some need in themselves? There are many anecdotes recounted that suggest it is so.

On a stormy night in a bleak spot in the north of England, a man was about to lock up and go to bed when he heard a gentle tapping on the window pane. It was an eerie sound so late at night, but he screwed up his courage, went to the window, opened it to look out and was swept aside by a rush of small birds. They were larks on migration. They swept into the room, settled on anything that offered a convenient perch and settled down for the night. By morning the storm had blown itself out. As soon as the owner of the house opened the window the birds started to take off and streamed out through the open window.

There is no way of checking this story. It may be a distorted or exaggerated account of what actually happened. It is, however, typical of many other stories from all parts of the world which suggest that some birds, in a moment of crisis, may use the beak to summon assistance, just as a child, in a moment of panic, may pummel a closed door without knowing beforehand there is someone on the other side who will help.

The third and smallest pile of letters relates to the behaviour of the Carrion crow of Europe. We are on more certain ground here, for in the south of England, some years ago, the owner of a house filmed what took place. He and his wife were awakened one morning at 5 a.m. by a confusion of sound outside the French windows of his bungalow.

Creeping round quietly to investigate, he found a crow flying repeatedly at the glass panes and crashing against them. Having collided with the glass, the crow would flutter down, turn round and return to the exact spot whence it had previously taken off. It would caw once or twice, then fly up, bashing against the window with its beak again and again as it fluttered down, clawing at the glass with its feet. For an hour or more it would repeat this, always with the ritual of returning to the same spot to take off, while its companion,

Image of an enemy
A man had twice noticed, when passing a particular spot, that he disturbed a chaffinch, as he thought feeding among fallen leaves in a gutter. Both times the bird merely flew to the top of a wall and waited for him to go by. The third time, however, having passed the chaffinch he stopped and turned to watch. The bird flew down and began a fierce onslaught on the polished wheel-disc of a parked car. He could hear the clink of beak and claws. This was by no means the first time a chromium wheel-disc has been used in this way, but it usually occurs early in the year. It shows how strong was the fighting spirit in the chaffinch that it made no attempt to fly away, only waited for the man to pass to renew the attacks. It is characteristic of these attacks that they are so persistent. When two male birds fight over a territory it is axiomatic that the one in possession fights more vigorously than the intruder, who soon accepts defeat and departs. The trouble with a reflection is that it stays, so long as its counterpart is there. Moreover, it shows just as much fight and refuses to give in.

264

presumably a female, stood around taking no more than desultory interest in the proceedings.

By the end of the hour, two whole panes, each seven feet by one and a half feet, and parts of the other two, were so thickly covered with splashes of blood and saliva that it was difficult to put a finger tip between them. Each day the owner of the bungalow cleaned the four large panes. Each following morning they were just as extensively sullied. This went on for over a week, at the end of which the attacks ceased.

There have been similar episodes reported all involving a Carrion crow and all following the same pattern. Perhaps the most remarkable part of the first of these occurrences was the persistence shown by the attacking crow. On the third morning the owner erected a barricade to stop the crow reaching the window. He placed a pair of step-ladders at each end of the French windows and ran planks between them on which he placed every empty wooden packing case or oil drum he could find. The crow had now to squeeze its way between them to reach the windows, but it still managed to repeat its attacks and continued to cover the glass with blood and saliva.

The loss of blood and body fluids must have placed a great strain on the crow but still it persisted, as if punch-drunk. This raises the question whether animals can show what amounts to **addiction.**

Among the familiar addictions to which human beings become victims are alcoholism, smoking and drug-taking. All three are sufficiently common among people of all races, civilized and primitive, that the question is inevitable as to whether they satisfy some deep, elemental craving common to all organisms, human or animal.

Alcoholism

It is not usual to find alcoholism among animals, but there are sporadic reports of birds showing the familiar symptoms of inebriation. Fruit-eating birds, such as thrushes, have been seen, after feasting heavily on over-ripe berries, to stagger slightly as they walked. Even more convincing are film sequences taken in recent years of large mammals in Africa, such as antelopes, walking away after having gorged themselves with over-ripe fallen fruits, lurching and zig-zagging, bleary-eyed, and showing all the familiar signs of being drunk. The most convincing of these film sequences showed an ostrich, crossing its legs as it walked, lurching, and with long neck wobbling as if 'drunk as a lord', after having filled itself with these fruits. And an elephant tipsy from the same cause is a sight to behold!

Perhaps 'addiction' is not applicable here, since alcoholism as such is dependent upon a steady supply of liquor and over-ripe fruits have a limited season. Even so, to all appearances the ingredients of an addiction are there, because the participants in these wildlife baccanalia will return again and again to the spot as long as the supply lasts.

Smoking

Smoking also must be a limited vice among animals for much the same reason, that the necessary materials are not readily available. There are indications, however, that but for this restriction a habit could be developed. Where domestic cows are being pastured it is often necessary to burn vegetable rubbish, such as hedge trimmings. As the smoke billows from the fire the cows gather round, walking over to the spot from several hundred yards if need be, and thrust their heads into the smoke, inhaling it.

Years ago, in the Johannesburg Municipal Gardens, South Africa, was a chimpanzee that apparently was given a lighted cigarette, which it proceeded to smoke. Its companion was also given lighted cigarettes and a third caught the habit from them. The evidence for this is in the South African Journal of Science for 1957. A fourth chimpanzee in the same cage did not develop the habit.

It is of interest that the chimpanzees are reported to have exhibited all the peculiarities associated with smoking in human beings. They would chain-smoke, if enough cigarettes were available, lighting a fresh cigarette from the burning stub of another. They would inhale the smoke, sit back with arms and legs crossed, exhale the smoke while twisting the lips and watch intently the patterns woven by the smoke. They showed the same intoxication with the first cigarette and, later, the same sedative effects as in human beings, and when the craving was on, would search desperately for an unused cigarette or for cigarette stubs with which to indulge. A chimpanzee would light a cigarette when handed a light. It recognized a cork tip and would turn the cigarette round to light the correct end.

Drug-taking

Drug-taking by animals is less easy to detect. Drugs are plant products, animals eat plants, and no systematic study has

Dog's time-sense

Many animals navigate by the sun and to do this they must be aware of the sun's movement. This method of steering was first discovered in ants, but it is now known to be widespread in insects. One consequence of it is that insects can instinctively calculate the hour of the day by the position of the sun or by the corresponding pattern of polarized light in a blue sky.

Cockroaches have a well marked daily rhythm, which is related to light and darkness (see p 26). It is governed by a clock in the suboesophageal ganglion, a knot of nerve-cells in the head lying beneath the gullet. A cockroach can be made to lose its rhythm, by keeping it under unnatural conditions of continuous light or continuous darkness. When the suboesophageal ganglion of a normal cockroach, with a well-marked rhythm, is cut out and transplanted into the head of a cockroach lacking a rhythm, the latter assumes the activity rhythm of the former.

When we come to higher animals we find a time-sense of a different order, as indicated by the many anecdotes of cats and dogs that seem to know the time of day as well as what day of the week it is.

One of the more amusing of these stories concerns a dog whose owner used to take it when he went driving. The dog's owner had the habit of stopping at a public house for a drink. He did this one day and ten minutes later the dog put its fore-paws on the driving wheel, with one paw on the button sounding the horn. The owner came running out to see what was wrong. Thereafter, the dog always sounded the horn ten minutes after his owner went in for a drink.

been made to link the two or to determine whether any animals are drug-addicts, *per se*. There was, however, a report in 1941, of a Grasshopper mouse in the New York Zoological Park that had the habit of chewing cigar stubs. Then it would part its fur with its forefeet and apply the chewed cigar leaf to its skin. It is interesting that the tobacco plant grows wild in the habitat of the Grasshopper mouse!

All this is anecdotal; it is not based on solid scientific foundations. Even so, the items given indicate possible lines of research or, what is more probable, fields of information in which an unexpected new discovery may suddenly pull the loose ends together and make a whole picture. It is surprising how many aspects of animal behaviour have been staring us in the face for a long time, and yet have remained unobserved until recent years.

Weather-forecasting

In all parts of the world, there are legends or sayings in which one animal or another is credited with forecasting the weather. The usual explanation by orthodox meteorologists is that animals only indicate what the weather is going to be when it is already upon us. However, even the meteorologists will admit that there are exceptions, though this is not to say that all folklore about animals and weather is correct. Certainly the implications of such names as the weatherfish, thunderworm, Rain frog and sunfish can be misleading. In recent years, there have been outstanding developments in the study of animals' foreknowledge of weather.

One bird that has been credited with foretelling storms is the Common swift, which is variously known as the rainswallow, stormswallow and devilbird because of the association of its movements with bad weather. It seems that swifts can detect the approach of electric storms while these are still a long way off. They respond by flying in a stream at right angles to the path of the storm, in extreme cases migrating hundreds of miles and being absent for several days, then returning after the storm has passed. When they do this during the breeding season their nestlings are left unfed and go into a temporary state of hibernation.

Shepherds can forecast local weather, on the short term at least, from the behaviour of their sheep, and in early 1975 there came a report that a farmer in Texas challenged meteorologists that he could forecast weather more accurately than they could from the behaviour of his pigs. It is claimed that during the period of the test he correctly forecast eight rainstorms out of ten, while the meteorologists forecast only one.

On a more scientific basis, it was revealed at about the same time that scientists in New Zealand had established that the movements of moths are an indication of weather to come. The surprise was that the moths were not influenced by humidity or temperature but by the ionization of the atmosphere (ions are electrically charged particles).

In many respects, animals' senses are more refined than ours; this is borne out by their sensitivity to earth tremors. Probably every animal is sensitive to vibrations in some form, although this sensitivity is more marked in some animals than in others. In some parts of the world, there are underground caves which contain large pools, or even small lakes, inhabited by blind fishes. These fishes are whitish or flesh-coloured and totally without eyes. They are, however, highly sensitive to vibrations in the substratum. Exactly what use they make of this sensitivity is still the subject of research. It may be used for avoiding obstacles and enemies or for finding food, guided by the vibrations set up by these in the water.

Even fully-sighted fishes, living in normal lakes, show a high degree of this particular sensitivity, as the following example will show. One windy day, as a woman was leaving her cottage, the wind slammed the door behind her. At the sound of the door slamming, she saw several fishes leap out of the water in the middle of a nearby lake. The same can be seen where heavy traffic passes near a lake's edge. This is not surprising, since vibrations in water are used by many kinds of fishes in their daily activities: in fighting, detection of enemies, and in courtship. It does come as a surprise to learn that many land animals are severely affected by earth tremors, before these are apparent to human inhabitants. Dogs whine, horses tremble, chickens panic, and birds fly around in an agitated way. The classic example of this occurred at Concepcion, in Chile, in 1835. At 10.30 a.m. the sky had been filled with screaming seabirds. At 11.30 a.m. all the dogs rushed out of the houses and horses showed evident signs of being uneasy. Not until 11.40 a.m. did the earthquake proper occur, which wrecked the town.

Overkill
Random observations of animals' movements often present puzzles which are difficult to unravel; this was particularly

Truth must out — eventually
It does not always follow that because generations of scientists have been sceptical of reports by laymen that these are necessarily incorrect.

Reports have been made again and again of cows being taken in to be milked that have been found to be 'dry', as if something had robbed them of their milk. Farmers blamed the hedgehog for this and over the centuries the belief became established that these animals were the culprits. Furthermore, farmers have testified to having seen hedgehogs in the act of sucking milk from the cow. A few have reported shooting such a hedgehog and finding its stomach full of milk. The general view expounded by competent naturalists has been that no doubt the hedgehog was searching under the udder of a reclining cow seeking insects attracted there by the warmth. Another often reported assertion has been that the hedgehog's gape is too small to allow it to take hold of a cow's teat.

On purely hypothetical grounds it has been possible to counter this scep-

There are occasional stories of small carnivores, such as weasels, killing more prey than they need and laying the carcases out in neat rows. These stories are treated with reserve by zoologists but there was one occasion when a domestic cat did precisely this. It found a rat's nest containing a number of half-grown rats at the far end of a garden. It killed the whole litter and methodically carried each one in turn a distance of about 20 feet, to the nearest clear space, on a bare path. There it laid the dead rats in a neat row, furnishing a clear example of overkill because it ate very little from the row of carcases. Even if it had been living in the wild and responsible for catching its own food, it would thus have killed far more than it could have consumed before the dead rats began to putrefy.

ticism by pointing out that a hedgehog's gape is one and a half inches, that hedgehogs have been seen to stand on their hindlegs and that in this position they would be within easy reach of the teats of a standing cow.

Within the last few years veterinary surgeons called to examine the injured teats of cows have declared that the injuries were the result of the teats being bitten lightly by hedgehogs. This opinion is reinforced by the fact, testified by several farmers, that cows only lose their milk during spring and summer, and this ties up with the fact that hedgehogs are in hibernation during most of the autumn and winter.

The habit is not widespread but very intermittent, which accounts for the long period needed to establish the truth.

true of the phenomenon known as overkill. From a long time back, naturalists have noted that a carnivore will kill more prey than it can eat. Indeed, it is not only naturalists who have been impressed by this. The familiar example of the fox in the hen-run has been noted as long as people have been keeping poultry. A fox that gets into a hen-run may kill a dozen chickens and carry off only one. This has been seen many times, with both the European Red fox and the American Red fox. A leopard has been known to kill seventeen goats at one time, leaving their carcases scattered around. Lions will kill more cattle, donkeys and goats than they can eat. A Polar bear was credited with killing twenty-one narwhals in a row, as they came up to breathe. Wolves will kill large numbers of caribou calves, domestic stock, or White-tailed deer. There have also been surplus killings of wildebeest calves by hyaenas.

Dr. Hans Kruuk, the Dutch zoologist, noted that the hyaenas' overkill of wildebeest calves occurred on stormy nights. He then turned his attention to the overkill of Blackheaded gulls by foxes. He found that on dark nights, with an overcast sky and no moon, gulls roosting on the ground made no attempt to escape when molested. Kruuk was able to pick them up without their trying to escape. He also found that a fox would walk through a gull colony, making no attempt to stalk the gulls, but simply biting one after the other, not always killing them but leaving some merely badly injured. This seems to be the pattern for other overkills.

The victims, whether killed outright or only injured, form a source of food for other carnivores or scavengers, but the underlying basis for an overkill lies in the behaviour of the prey. The weather and other conditions have changed the pattern of their behaviour, inhibiting their escape reactions. When chickens are enclosed in a run, it is the fact that they are unable to escape that leads to overkill. Weather then has nothing to do with it.

Objectively, overkilling is no more than a changed pattern of behaviour in relation to particular environmental circumstances, in these instances an overcast sky and the number of days to a new moon. Earlier writers, and some 'nature writers' still today, use lurid terms to describe such behaviour. They often refer to carnivores, especially the smaller ones, such as weasels and ferrets, as 'blood-thirsty killers, seeking whom they may devour'. The modern approach is to deal with such matters as the sum total of a behaviour pattern, which takes the sting out of it.

Opportunism
Another favourite term used to describe certain kinds of animal behaviour is opportunism, a word which implies that the animal has, by some process of intelligent reaction, seized upon particular circumstances which are greatly to its own advantage. More soberly, we can interpret opportunistic behaviour, as we now interpret overkill, as a change in the natural pattern of behaviour which has resulted from a change in environmental conditions.

A typical example would be when an animal which normally hunts for a particular item of food, turns to one that is more abundant but less familiar, when its normal food is scarce. The Common rat normally eats grain but will find sustenance in almost any food, plant or animal.

Cannibalism
This is the moment when a word about cannibalism would not be out of place. Cannibalism is the eating of other members of the same species; it is seen in both animals and

Cannibalistic octupuses
Octopuses are molluscs, relatives of snails, clams and oysters yet remarkably unlike them. Not only do they, in some species, run to giant size but their swimming abilities and consequent vigorous movements are in sharp contrast to the sluggish locomotion of typical molluscs. They also have remarkably large and efficient eyes, comparable to those of vertebrates and they have a far greater intelligence than the great majority of molluscs. These things, and in particular their large eyes and their eight writhing, sucker be-studded arms, give them a decidedly sinister appearance.

If one more feature were needed to put the seal on their evil reputation it is found in their strong tendency towards cannibalism. In this, it is usually a large octopus that eats a smaller one, but not always. From captive octopuses, kept in marine aquaria, we learn that even when fully fed an octopus is prone to devour another octopus, not wholly, because the rule seems to be that they bite off and devour the arms of their victims. One, for example, ate five of the arms of a smaller octopus and then dropped it, the victim succumbing two days later.

Octopuses occasionally make meals of parts of themselves, a process known as **autophagy**. One such ate five of its arms and suffered fatal consequences.

Patient amputation
Starfishes can penetrate the fortress-like shells of oysters and clams, but they fall victim to a shrimp. It is a very odd shrimp, at most two inches long, the male being smaller than the female, and coloured a garish white with blue patches. The shrimp was unknown until 1852, when an American expedition to the South Pacific found it. It has since been found from Hawaii to Australia and across the Indian Ocean to East Africa.
Known as the Painted shrimp or Elegant coral shrimp, it is rarely seen because it hides under coral rock by day. When hunting, male and female approach a starfish and begin to nibble at the base of one of its arms, in slow motion. Surprisingly, the starfish makes no attempt to throw them off. Perhaps because, if it loses an arm it can grow a new one.
Over two to three days the shrimps nibble away until the starfish arm is severed. Only then does the starfish move away leaving the severed arm, still alive, for the shrimps to continue their slow-motion feeding.

man. Even the lowly, single-celled amoeba has been seen in the act of eating a smaller amoeba. Insects, crabs, scorpions and spiders, fishes, frogs, reptiles, birds and mammals have all been guilty of it. Among birds an outstanding example is the Herring gull, which is frequently seen killing and eating the young of its own species. Among most animals, it is the young that suffer. Lions, among mammals, sometimes kill and eat each other. Among insects mainly eggs are eaten; examples of this are found in the caterpillars of the Monarch and Queen butterflies.

Civilized people are revolted by the idea of cannibalism, whether in animals or man. However, looked at objectively, cannibalism can be seen, with rare exceptions, as the result of overcrowding (when it becomes an instrument of population control), of feeding frenzy (see p 75), or of psychological stress (when female voles, mink and other mammals will eat their young). Rarely is it the result of shortage of food. Just as overkill must be exceptional, since no species can afford to wipe out its prey species, so cannibalism must be exceptional if a species is not to cause its own extinction.

Sibling murder
There is, however, a refined form of cannibalism, known as sibling murder. Siblings, properly speaking, are the young of a parent or parents born at different times, in contrast to twins or triplets, which have developed together in the womb. Strictly speaking, two young birds hatched in the same clutch are twins, and three are triplets, but a distinction is made in the case of birds-of-prey, both hawks and owls. Birds-of-prey lay their eggs at intervals of two, three or more days, and since incubation usually starts with the first egg laid, the chicks hatch at different times. Where, as in smaller birds-of-prey, the clutch is larger than three, incubation may end with some of the eggs not hatching. Mostly, with rare exceptions, the chicks are of different sizes, and if food is short the first-hatched will kill the other or others, in what has aptly been called a Cain and Abel battle, or, less elegantly, sibling murder. This is a safety device. If parents have difficulty in finding food, the smaller chicks die and the remainder will be fed and so have a much better chance of survival.

It is inevitable that in any account of the ways of life of animals, much has to be said about killing and injuring. To the civilized human being, such topics are often abhorrent. From the cradle to the grave, especially in advanced civilizations, the emphasis for human beings is on security. We grieve, if only momentarily, on seeing a baby bird fallen from the nest. Yet the fact is that among birds 60-75% of all young perish before reaching three to six months of age, the figures varying with the species. Comparable rates of mortality, especially in infancy and certainly between birth and sexual maturity, could be given for all kinds of animals. Only in this way can populations be kept stable.

Pain

The distress experienced by many people at seeing animals killed by predators can be obviated to a large extent by an understanding of what pain is involved. How much do animals feel pain? This question is often asked. In general, we can be sure that the capacity to feel pain diminishes as we descend the animal scale, and the questioner is usually concerned only with higher animals. One looks therefore for pointers that will give guidance. There is the classic, if grisly story of the fish being hooked by an angler. The fish escaped but left its eye impaled on the hook. The angler recast his line, with the eye still on the hook and in a short while reeled in his line only to find that the injured fish had accepted the bait of its own eye. One knows, from many written accounts as well as from firsthand observation, that some animals deal out savage punishment to their fellows in territorial fights, and that these injuries neither shorten the life of the victim, nor seem to affect its well-being; its readiness to fight again seems to be unimpaired.

Hippopotamuses, especially males, engage in truly spectacular fights in the water. They slash at each other with their enormous tusks, rearing up amid a great turmoil and turbulence of water to do so, and inflicting long gashes on each others' flanks. An occasional outcome is that one of the protagonists has a foreleg broken. This is fatal because the weight of the huge body allows no chance for the limb to heal. Otherwise, so we are assured, the wounded hippo squeals with pain for a while as water enters the wound, but healing is rapid.

Once a predator has seized its prey, pain seems to be non-existent, judging from the behaviour of the prey, which often seem to be hypnotized. Several men who have been seized by a lion and been rescued have left written testimony to this effect. That set on record by David Livingstone, the famous explorer, is typical: 'The lion growled horribly in my ear and shook me like a terrier does a rat. The shock

272

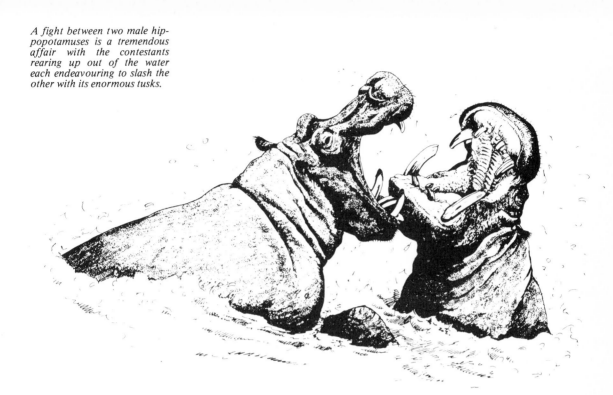

A fight between two male hippopotamuses is a tremendous affair with the contestants rearing up out of the water each endeavouring to slash the other with its enormous tusks.

In the same area two men were staying overnight in a deserted shack. They placed a lighted candle on a shelf and to their surprise it started to move up towards the roof. A packrat was carrying it away.

Many of the smaller rodents are noted for hoarding, not only food but other objects also. The packrat is unique in swapping. Presumably it is carrying something away, comes across an object which from its colour or brightness attracts it more, so it drops the thing it was carrying and picks up the new object. It is difficult to believe that packrats act on the principle of fair exchange being no robbery.

produced a stupefaction in me, like that which a mouse must feel when caught by a cat. It induced a sort of state of anaesthesia, in which I felt neither the pain nor the shock, although I was fully conscious at the time. I was like a patient slightly under the influence of chloroform, who sees all the motions of the operation he is undergoing yet can feel no trace of the knife. This extraordinary state was not the outcome of any mental process, but the shock removed all trace of fear, and eliminated all horror — even in the very face of the lion.' The state described by Livingstone is probably related to the phenomenon known as **akinesis**.

Akinesis

In 1646, Father Athanasius Kircher performed his *Experimentum Mirabile*. He tied together the legs of a partridge, put the bird down quickly on its back and drew a chalk mark from its beak forwards. The partridge lay still, as if dead.

The 'wonderful experiment' has been repeated many times since then, on all manner of animals, from insects to cats and dogs. The partridge, as these other experiments have shown, did not need to have its legs tied, nor was there need of chalk, to hypnotize it. Merely putting it suddenly and gently on its back would have been enough. This was demonstrated by Ernst Mangold, who made an apparatus for quickly and gently turning an animal on its back. With it he hypnotized many young animals (see p 171).

Another well-known trick is to turn a crayfish upside-down so that it stands on its head and two claws, the so-called tripod position. The crayfish will stay motionless.

If the Egyptian cobra is picked up, held behind the head, and the forefinger pressed into the top of its head, it becomes quite still; if it is laid on the ground it looks like a lifeless strip of material. It can be awakened by blowing at it or flicking water over it. A frog or toad, suddenly turned onto its back, can be lifted up by one leg without rousing it.

Scientists prefer to call this sudden lifeless state akinesis. It is very like the 'shamming dead' used by such animals as the American opossum, a state scientifically known as **thanatosis.** Both have similarities to hypnosis, as practised medically on man, except that there is no question of speaking or giving orders to hypnotized animals. Their akinesis or thanatosis is a response to sudden change of position, shock or fear. The practical result with animals is that they may escape danger. Predators such as wolves, cats (large and small) and dogs tend to lose their hunting instinct temporarily at the sight of a non-moving prey.

Curiosity

It has long been recognized that there are three compelling, basic motivations in every living organism, the motivations to feed, to escape enemies, and to reproduce. A few years ago it was suggested that there is a fourth motivation, a sense of curiosity. It was further suggested that this was the strongest of the four. Curiosity leads to the exploration of the environment by young animals, without which they would not become independent. After the animal has become independent, exploration or exploratory behaviour (see p 30) is necessary to find a nest site, lair or den, to find food and to find a mate. A sense of curiosity is nothing unusual, but it can lead into unusual channels, as it does in the behaviour known as charming.

Charming

There is an age-old saying that curiosity killed the cat, but more animals than *Felis catus* have been led to disaster by it, if we are to believe the many stories told of beasts-of-prey

Story of a dead duck
A few years ago someone found a duck buried alive in his rose bed. This was in England. A fox was suspected but there the mystery remained until two scientists published an account of their observations, made in the same year in the United States.
They found that a Red fox always attacks a duck from the side or from behind. The duck tries to escape. If the duck sees the fox and turns to face it, the fox backs away, especially if the duck walks towards it.
Should the fox steal up on a duck and touch it, with its mouth or paw, the duck makes no attempt to escape. It goes into a sort of hypnosis, or death feigning. Its neck and head are stretched forward to the full, in line with the body. The wings are held tight to the body. The duck remains rigid with its legs and feet stretched out behind, but its eyes remain open.
The duck may hold this position for as much as a quarter of an hour and while it does so the chances are the fox may not bite, or bite only lightly. The fox may even cache the 'carcase', digging a shallow hole with its forepaws and throwing earth lightly over it with its nose-tip.
Should the fox wander away the duck may slowly raise its head and look around. If it again sees the fox it will slowly lower its head again to take up the 'dead' position. If the fox is out of sight the duck quickly takes wing and flies away.

charming their intended victims. A typical account runs as follows. A fox or stoat is seen behaving as if it had taken leave of its senses. It prances about, bucks on stiff legs, rolls on the ground and somersaults, all on one spot. One by one, birds and rabbits in the vicinity leave what they are doing and fly or walk over to gather in a rough circle to watch the antics. Suddenly, the fox or stoat ceases its lunatic activities and pounces, so getting a free meal.

Similar stories are told of other members of the weasel family; the best authenticated account is about a crow seen on a hilltop in southern Asia. It was walking over the rocky ground to where a Himalayan marten was rolling, tumbling and somersaulting on the ground. It was then seen that another marten, presumably the mate of the one performing the acrobatics, was stealthily ascending a slope in a wide détour towards the crow, unobserved by it. Finally, it made a dash and seized the crow, whereupon the first marten ceased its clowning and joined in the kill.

One thing of which we can be certain is that birds will be attracted to go over and watch an animal such as a squirrel playing on the ground. There is, however, some doubt

Stratagem used by a pair of Himalayan martens, in which one held the attention of a crow by its seemingly playful antics while its partner was observed creeping up on the bird to capture it.

whether charming is a real phenomenon and not just a matter of coincidence. Foxes and weasels will often play on their own, especially in country where they are not disturbed. It could follow naturally that birds and small mammals might gather to watch, and that the fox, weasel or marten then takes advantage of the situation.

Then comes the question — what would start the fox, weasel or marten on this kind of behaviour? We can never be sure that there is not another cause. The domestic cat can sometimes be seen behaving in a way reminiscent of the alleged charming. It has long been known that members of the cat family seem to go crazy over the plant known as catmint in Britain and catnip in the United States and Canada. They roll about in a bed of catmint and rub their faces on the ground.

'Catmint' is Mediaeval English. The plant itself is known in medical Latin as *Nepeta cattaria.* In Italian it is known as *herbaddei gatti,* in French as *herbe aux chats,* in German as *Katzenminze,* all of which mean much the same thing. The similarity of its name in so many European languages suggests that its effect on cats has been known for centuries.

In spite of this, however, experiments have shown that essence of catnip actually has a soothing effect on the nervous system of cats, and it has been used in many circuses for gentling the large cats. In North America it has been used as a lure to bring pumas and bobcats within gunshot or into traps. It follows, therefore, that any unusual antics shown by these animals in its presence were due to a pleasurable sensation received from it.

Nobody could make sense of this until two Canadian scientists investigated the phenomenon. They found that the movements made by a cat when confronted with catmint were comparable with those of a she-cat or queen when coming into season. The cat in season goes further, however, doing much sniffing and licking in addition to rolling. She will make a pass at any cat or cat-sized animal that goes near her. So what a cat does when it sniffs catmint is only a small part of the pattern of pre-mating behaviour in the queen.

Only half of the dozens of cats used in the investigation reacted to the catnip essence. The experimental group included kittens, adult queens and toms, normal as well as neutered or spayed. So the reaction is independent of age or sex, or of whether the cat is entire or has been doctored. The only conclusion is that some cats react to a particular odour by displaying a small part of a normal pattern of behaviour,

Most birds sing more vigorously during fine weather than during inclement weather. One exception is the Mistle thrush of Europe which has the alternative name of stormcock and can sometimes be seen in a snowstorm or a partial blizzard clinging tightly to the topmost twigs of a tree and singing its heart out as its perch is whipped backwards and forwards by the wind.

276

Birds' balancing act
Why is it birds can stand or lean on one leg? The question has often been asked. When a man tries to do this he finds it difficult and it is largely a matter of weight upsetting a centre of gravity. The human leg forms a fair proportion of the total weight of the whole body. So even when one makes an adjustment to shift the centre of gravity the leg is a heavy drag on the supporting muscles.

By contrast a bird's leg is slender without significant weight, and it can be drawn right up under the body. Added to this the long toes, well splayed out, more effectively spread the load than do the human feet.

There are times when one can get close enough to a small bird perched on a branch or a post and by keeping very still watch what is happening. It is often possible to see that the claws on each toe are acting as grapnels in the tiny irregularities in the surface of the bark or wood. A bird on one leg is as steady as a rock even in a strong blustery wind which ruffles its feathers. Its balance even then is not upset and the efficiency of the claws must be a large part of the answer.

and that the actions they then perform fulfil no purpose. It probably means that the chemical composition of the catnip essence must be near that of the pheromone (at present unidentified) which touches off the complete sequence.

This is more important than at first appears, when dealing with the more unusual behaviour, especially that found in the higher animals. Cats, for example, are said to rush madly about the house when rain is approaching. In this, as in many other puzzling things animals do, it is conceivable that they are reacting to some strange odour, sight, sound, change in barometric pressure, or other circumstance at which we cannot even guess. Should any such circumstance stimulate the appropriate sense-organs, perhaps only incompletely, so that the messages transmitted by them to the brain and thence to the muscles give rise to an incomplete series of actions, these actions would appear to us quite purposeless. Yet they would really be actions out of context, which were but part of a larger, normal pattern.

Aberrant instinct

Some animals do things that seem to be instinctive, yet which are almost incredible. Because they seem to be a perversion, the term aberrant instinct has been used for them; the implication is that this is an instinct which has developed along the wrong lines although the very survival of the species shows this is not the case. One insect to which the term is applied is the Large blue butterfly, of Europe.

For a long time naturalists were puzzled that they could find no caterpillars of the Large blue butterfly. There were plenty of adult butterflies. Then the reason was discovered. In June, the females lay their eggs on wild thyme, and ten days later the caterpillars emerge. For three weeks they feed on the flowers of the thyme, and they also devour the smaller larvae of their own kind when these are in the process of emerging — an act of cannibalism.

After the third moult the remaining caterpillars drop to the ground and wander aimlessly until discovered by a Red ant. Instinctively the ant recognizes that the caterpillar exudes drops of a honey-like substance that the ant enjoys. The ant strokes the gland exuding this a few times, sips up the syrup, then takes up the caterpillar in its jaws and carries it off to its nest. In the nest the caterpillar feeds on the ant larvae and grows quickly, the adult ants frequently milking it for its syrup. During the winter the caterpillar hibernates, completing its growth the following spring, when it pupates.

277

Three weeks after this the adult emerges and crawls off.

The butterfly seems to have become parasitic on the ants and is apparently unable to survive without them. More remarkable, it has been found in recent years that the ants become so addicted to the sweet substance given off by the caterpillars that once they have carried them to their nests they not only allow the caterpillar to feed on their own larvae, but also neglect the larvae themselves. In time the colony harbouring the caterpillars tends to degenerate.

Fossil behaviour

'Aberrant instinct' is a term that seldom appears in the natural history literature; another such term is fossil behaviour. Both are worth keeping in mind as possible explanations of otherwise puzzling behaviour. Just as in our bodies and in almost any body of a higher animal, there are structures that are now almost obsolete, so there seem to be behavioural traits that belong to the past. One such structure we can think of is the appendix, although that is not so much obsolete as an organ that has changed its function. Another is the human tail. In most of us, the tail bones are hidden under the skin and superficial muscles, but occasionally one comes across a person with an inch or two of tail still protruding. It is logical to suppose that there are also tricks of behaviour that once served an important function in the ancestors of an animal species and are now seen in their descendants, perhaps infrequently, as a sort of relic. Such traits, if they occur, are unusual and usually intermittent — they are difficult to identify or interpret. A familiar example is that of the dog turning round and round before curling up to sleep on a level carpet. It is always assumed, and reasonably so, that this harks back to the time when its wild ancestors rotated to trample long grass before dropping to the ground to rest. Another is suggested for the behaviour of the platypus. The female of this Australian egg-laying mammal digs a tunnel with a chamber at the end in which to lay her eggs and rear her young. Along the tunnel she builds a series of mud walls which would make entry difficult for a predator. The platypus now living has no enemies except for the very occasional hazards of visits by snakes or goannas (monitor lizards). It is presumed therefore that the female platypus is continuing a piece of behaviour essential to its ancestors against enemies that have died out.

The spawning migrations of eels (pp 197-198) could also be regarded as a form of fossil behaviour.

Ants transporting a caterpillar of the Large blue butterfly to their nest where the workers will sample the sweet substance the caterpillar gives off, while the caterpillar itself will feed on the ant larvae. This extraordinary story constitutes one of the minor mysteries of the insect world.

Chapter Thirteen
Unsolved Problems

anting in birds — ant-bathing — smoke-bathing — fire-bathing —
self-anointing (hedgehogs) — cairn-building — blood-squirting —
autohemorrhage — dawn chorus — rat legends — porpoising otters
— compassion — compassionate killing — rooks' parliament —
dreaming and recalling — telepathy — homing snails — magpies'
mass meetings — animal artists

Anting

When people first started talking about having seen birds 'anting' they were ridiculed by the ornithologists. There came a time, however, less than half a century ago, when 'anting' was shown to be real. Today, no ornithologist feels his education is complete until he has seen a bird ant.

Put simply, the peculiar behaviour we call anting happens as follows. A bird picks up an ant in its beak, at the same time extending one wing (some species spread both wings). It appears to rub the ant on the undersides of the flight feathers of the spread wing. Then it either casts the ant aside or swallows it, picks up another and repeats the performance, either under the same wing or under the other.

Anting has been seen performed by more than two hundred species, all songbirds (Passeriformes). Among individuals, some birds seem never to ant, while others ant almost at the drop of a hat. Young birds only three days off the nest have been seen to ant. Aviary birds have been known to live up to eighteen years before anting. As with smoking, and other such addictions, there are the habituals, the moderates and the abstainers, as well as those that try it and give it up and those that take to it only in old age.

Anting takes different forms according to the species. The typical posture is that shown by the starling, top right, in which the bird half extends both wings and rubs an ant it has picked up on its beak on the underside of the wing. Some species extend only the wing under which the ant is being rubbed. A peculiarity of anting in the Common jay of Europe, top left, is that it ants without picking up an ant in the bill. It is stimulated to do this merely at the sight of ants and the manner in which its wing primaries are spread allows ants to crawl up over the wings, even on to the body and the head. The jay's method is to that extent halfway between typical anting and ant-bathing (p 289). In the lower picture a European magpie is removing ants from its plumage after an anting session.

Above: Alone among the cat family, the lion is a social animal. Perhaps because it is a slow runner, it has to cooperate with other lions to hunt successfully. This lioness and her cubs are part of a pride comprising up to three dozen lions. The two or three male lions have little to do with family life but they protect the cubs from neighbouring lions. Hunting is a communal activity, several lionesses together having more success than one on its own at stalking large prey such as giraffe and zebra. The males do little hunting but take the first serving at the kill and the cubs may starve if game is scarce.

Overleaf: Baboons of the open savannah survive by their strict social order. There is a rigid dominance hierarchy among the males and on the move the troop assumes a certain order. These Olive baboons of Uganda, however, live in woods and their arboreal life is more easy going. The hierarchy is more relaxed, and there is no order of march. Harmony between all members is promoted by their habit of grooming one another's fur.

The Tree hyrax, a small East African mammal surprisingly related to the elephants, is responsible for a spine-chilling nocturnal chorus. Each hyrax gives a series of screams and is answered by others up to a mile distant. The significance of these calls is not known but they may be between rival males. Tree hyraxes in one forest can sound very different from those in another part of the animal's range. This species calls in a series of screams descending the scale, ending with a gurgle as if the hyrax were being throttled. Other species sound like babies crying or have ascending screams. Tree hyraxes also bark a warning of approaching danger.

Right: The male Groove-crowned bullfrog calls loudly from a large puddle. The deafening volume of sound is produced by the frog shuttling air between the lungs and the two bluish-white airsacs and across the vocal chords. The airsacs puff out at each call. By calling from water, the bullfrogs attract females to the spawning grounds but the calls also serve to keep the rival males apart and, in one species of frog at least, they enable the female frogs to tell which is the dominant male.

Below: An African Secretary bird playing with a tuft of grass, throwing it into the air with its feet and leaping after it. Play has been rarely recorded in birds but there are instances of birds apparently frolicking just for fun. The Secretary bird may be well fed and passing the time by playfully attacking the tufts of grass. Although light-hearted, such play has the serious function of perfecting hunting skills.

Opportunists in the animal world make use of new features of their environment. On Barbados several acrobatic species of birds have taken to using dripping pipes and taps for drinking and bathing. This Lesser Antillean bullfinch hangs upside down and lets water trickle over its plumage and into its bill while it flutters its wings. This is just one small example of how birds have come to exploit man-made situations. Swallows perch on telegraph wires, martins nest under eaves and tits open milk bottles to steal the cream.

Right: Wood ants dragging a small tortoiseshell to their nest give an impression of intelligent cooperation but it is easy to credit animals with too much intelligence. Each ant is impelled to drag the butterfly towards the nest so their individual efforts have the effect of a coordinated aim and progress is made.

Below: The strange habit of firebathing. This tame rook has learned to light non-safety matches by striking them with its bill. Then it spreads its wings over the flame and smoke and holds the smouldering matchend to its feathers. The rook gets very excited during the performance but it hardly singes its feathers. Firebathing is very like the equally odd habit of anting (see text).

A hedgehog self-anoints by licking and chewing an object until it has worked up a foamy saliva, which it transfers to its spines by repeated flicks of the tongue. Occurrence of self-anointing in captive hedgehogs is erratic but there is a suggestion that the practice may be a means of disseminating the hedgehog's scent to other hedgehogs. As often happens, it is difficult to interpret a facet of an animal's behaviour until we know more about its private life and, meanwhile, self-anointing remains a mystery.

Snail boomerangs

In 1881, the scientific journal *American Naturalist* carried a story of a young girl who trained a snail to come to her call. It seems, however, that once the snail had left its hiding place it would withdraw into its shell on hearing a strange voice but would come out again if the girl called it.

This is a remarkable story because snails have no known organs of hearing. They have pits in the skeleton that act as organs of balance, and organs of balance in other animals are associated with the organs of hearing. Presumably there is an outside chance this story could be true.

More readily credible is the story of a man who found a snail on his precious pot plant standing in a balcony. He threw the snail to the ground, a distance of 20 feet. The next day the snail was back and it returned as often as he threw it to the ground — and it had to climb stairs to do so.

We know snails will, and often do, climb, and we know they home. A simple experiment to initiate children into the ways of biological research is to set them collecting snails. They dab the shell of each snail with paint, then throw the snails to all points of the compass and to distances of 20 feet or so. By the next day all the snails, other than those eaten on the way by birds or other animals, will be back where they began.

Like a boomerang, a snail hurled through the air will return to the starting point, but not so fast as a boomerang. It has long been said snails home using a sense of smell. If true this would be remarkable. A more recent suggestion is that they use the earth's magnetism to assist them on the way home. This, if true, would be even more remarkable.

It is a fantastic sight to see a bird anting. The actions are carried out at high speed, with an air of intense excitement, the bird often throwing itself into contortions, even toppling over.

The process is infectious. Often, when one bird ants, those around will also do so including individuals standing where there are no ants. These merely go through the motions, but lack nothing in excitement or the energy of the performance.

The most obvious explanation of anting is that it is a means of using the formic acid given out by ants for dealing with vermin in the plumage. However, there is little to support this idea and a great deal to be said against it. There have been many other theories to account for the habit, none very convincing, a few ridiculous and the rest difficult to substantiate. For some years anting was regarded as the greatest ornithological mystery. Today, the currently accepted view is that it is a natural method of toning the feathers.

This suggested therapeutic value of anting must, nevertheless, be questioned. A long series of observations carried out on aviary birds, and supported by observations on wild birds, has shown that an anting bird always starts with the left wing and, over a period, anoints its feathers with formic acid three times on the left side for every once on the right. This being so, it seems plausible to say that if anting tones the feathers, anting birds should be more glossy on the left side than on the right. It would then follow as a necessary corollary that a bird that habitually ants should tend to fly in circles, since the flight feathers on the left side will be stronger and in better condition.

A variation of anting is known as ant-bathing. It is a habit most commonly seen in members of the crow family. In this variation, the bird settles breast to the ground, over an ants' nest, spreads it wings, partially fluffs its body feathers, closes its nictitating membrane (third eyelid) and allows the ants to swarm over its body and wings.

Smoke-bathing and fire-bathing

The mystery of anting is not decreased by two variants of it, known respectively as smoke-bathing and fire-bathing. Usually birds, in common with other land animals, are justifiably terrified of fire and smoke. There are exceptions, however, again most commonly among members of the crow family. Starlings, which are as 'addicted' to anting as crows, are equally given to smoke-bathing.

In a typical example of smoke-bathing, one or more crows or starlings will be seen on the rim of a chimney from which smoke is issuing, with their wings spread as in anting. From time to time they will, so to speak, take a beakful of smoke and seem to place it on the underside of one wing, starting with the left wing and placing smoke three times as often under the left wing as under the right. A smoke-bathing session, provided the birds are undisturbed, may last for several minutes on end.

Fire-bathing is less common than smoke-bathing, yet in the sixteenth and seventeenth centuries especially, in England, certain birds were known as *aves incendiaria*, because of their habit of flying with glowing embers in their beaks onto the thatched roofs of houses and setting them on fire. These fire-birds included choughs, rooks, crows, jackdaws and magpies, all members of the crow family. There have been several well-authenticated cases of such fire-birds in recent years. The best documented case was that of a Carrion crow which was hand-reared from a fledgling and lived twenty years in an aviary. It would fly to a heap of burning materials and behave towards the flames as other birds behave in the presence of ants or smoke, either lying breast to the ground with wings spread or standing over it with spread wings. From time to time it would bite at a tongue of flame and appear to place a non-existent beakful of flame under one wing, always starting with the left wing and performing the motion three times, on average, under the left wing to once under the right wing.

This proclivity on the part of the crow for playing with fire was discovered accidentally and was exploited experimentally so that in its twenty years the crow fire-bathed hundreds of times. The usual procedure was to put a small heap of straw in its aviary and light it. The operation was difficult because as fast as a match was struck the excited crow would snatch it from one's fingers to hold it under its wing.

No cruelty was involved and the bird managed to perform without ever burning its feathers. Its eyes were protected because, as in anting and smoke-bathing, the nictitating membrane was always drawn across the eyes. The mouth was protected because, as in anting, there was always a copious flow of saliva. The feathers were protected because the movements of the wings fanned the flames away from the body, so that the bird had to stretch its neck to reach out to snatch at the tongues of flame.

Birds can sometimes be seen perched on the rims of chimneys when these are issuing smoke. Starlings and rooks are chiefly involved. As the smoke curls round them they spread their wings as in anting and the whole behaviour suggests that they are anting in the presence of smoke.

Mystery of the magpies' mass meetings
European Common magpies usually go about in ones or twos, but during January and early February they sometimes gather in groups. There may be six, 20 or 200. The purpose of these ceremonial assemblies is something of a puzzle to the experts. It looks like a mass courtship except that the birds are already paired.
There is much flying and chasing, on the ground or in the trees, with the cock birds displaying and posturing to the hens. An essential part of the ceremony seems to be the presenting of gifts, usually sticks, presumably symbolic of nest-building. The same meeting places are said to be used

year after year, and sometimes at times other than winter, when for example, a bird loses its mate during the breeding season. Presumably the gathering consists of non-breeding birds, which make up a percentage of the population.

The assemblies are interpreted as social gatherings, not specially for breeding. All the same, there is an air of great excitement. Crests are constantly raised, tails lifted and opened and closed like a fan, and the white body feathers may be fluffed out almost obscuring the darker patches.

The similarity of bodily and wing movements, the actions, the postures, the seeming excitement and the total pattern of apparent addiction displayed in anting, ant-bathing, smoke-bathing and fire-bathing are so alike as to leave no room for doubt that all are related phenomena. If, therefore, anting is to be supposed to have a functional value, then the other three must be considered to be due to an aberrant instinct. In view of what has been said, it is difficult to believe that even anting is functional, although there are ornithologists prepared to distort the logic of the situation to prove otherwise. So the mystery still remains: are these related phenomena functional or due to an aberrant instinct (p 277) or an addiction? Are they fossil behaviour (p 278), something that once had function? You toss a coin to decide!

Self-anointing

Another unsolved problem that bears a close comparison with anting is the self-anointing of hedgehogs. This was first described in detail by Konrad Herter in a monograph on the hedgehog published in German in 1938, although nothing appeared on the subject in the English literature until one of the present authors described it in the *Illustrated London News* of October 29, 1955, and coined the term 'self-anointing'. Now it has become well known and, as usual, many theories have been generated to account for it, none of which has so far proved convincing.

In a typical example of this behaviour, a hedgehog will be seen licking something, usually an object or material with a strong smell, although the range of substances so far recorded is very wide. The animal licks persistently at one spot while a foamy saliva accumulates in its mouth. Then it turns its head to one side, contorts the body, throws out its tongue and deposits the saliva on the spines of its back or flank. It then returns to licking the same spot and repeats the process of depositing the saliva on the spines, usually on another part of the body. A self-anointing session may last twenty minutes or more, at the end of which the hedgehog's coat of spines is an unpleasant mess of spittle.

Self-anointing has, for obvious reasons, been more commonly seen in pet hedgehogs or those kept in captivity for study purposes. Wild hedgehogs also do it since the dried spittle can sometimes be detected on their spines. There is the same pattern of frequency as in anting. Among pet and incarcerated hedgehogs at least, some are habitual self-anointers, others are never observed self-anointing, some start young (before weaning), and others take to it in old age. Both males and females have been seen self-anointing. A litter of half-grown hedgehogs once seen, all self-anointing at once, presented what must have been one of the oddest sights in natural history.

The theories advanced to account for self-anointing are various: it has been called a form of grooming, a displacement activity (see p 56), a means of getting rid of skin parasites, a method of masking the natural body odour to prevent detection by predators, and a pre-mating activity. On the scanty evidence so far produced the most promising theory is that it may have something to do with sexual activity; wild hedgehogs appear to self-anoint only during the breeding season. On the other hand, some hedgehogs seem to self-anoint with anything strange they may encounter.

Tobogganing tortoise

Tortoises are open-air creatures. They live hum-drum lives, usually where the ground is flat or only slightly undulating and the last thing we expect to see in a tortoise's habitat is a flight of stairs.

A man had a pet tortoise about a foot long which he kept in his flat on the first floor. From time to time he would carry the tortoise downstairs and into the garden, putting it on the ground to take exercise. Always the tortoise made for one corner of the garden and there settled down to rest.

One day the tortoise was missing from the flat. After an extensive search it was found in its favourite corner of the garden. A short time after this the owner of the tortoise saw it go to the head of the stairs and cur-ious to see what would happen he sat and watched. The tortoise pulled in its head and fore-legs, gave a push with its hindlegs and tobogganed down the 20 steps. Arriving at the bottom it placidly made its way to its favour-ite spot in the garden.

Perhaps the most astonishing feature of this spectacular phenomenon is how long it escaped notice by zoologists. Gipsies, who traditionally roast and eat hedgehogs, have known of it for a long time. People who keep a pet hedgehog have seen it; one described her pet hedgehog as having the habit of 'licking one of the legs of the sideboard and then going into convulsions'. One experienced naturalist admitted, when told about it for the first time, that he had often seen it but had not given it a second thought.

Self-anointing, anting and similar unusual and intermittent, almost random, behaviour are difficult to explain because they seem not to follow known rules and are difficult to study by repeated experiment. The main hope of finding an explanation lies in 'discovery by accident'. Cairn-building by the European Long-tailed fieldmouse is even more sporadic and difficult to elucidate.

Cairn-building

The Long-tailed fieldmouse burrows in the ground in fields, gardens and woods. It is very numerous and the round holes, little more than an inch in diameter, are equally so. Very occasionally, perhaps for one in a thousand or one in ten thousand, the entrance to the burrow is marked by a heap of small stones. Usually this takes the form of a pile of pebbles standing beside the entrance to the burrow; less commonly, the pebbles are taken into the burrow and accumulated near its mouth. Rarely, the cairn is built over the mouth, a sort of stone igloo with an opening to one side, as if intended to protect the burrow from rain.

The pebbles, some as heavy as the mouse itself, are collected from within a radius of up to six feet. The cairn remains there for a while and the pebbles are then scattered, by the mouse, all around. Even so, if the pebbles are removed by human hands and placed several yards away from the mouth of the burrow, the mouse will carry them back one by one to replace them on the spot whence they were taken — and later scatter them on its own accord.

The best one can suggest at this moment is that since the Long-tailed fieldmouse is a notorious hoarder of nuts, acorns and berries, this may be a perverted form of hoarding. However, the normal hoards are usually accumulated under the surface of the soil. This is true for rodents in general and one of them, the packrat of North America, will steal bright objects to hoard, and for every one taken it will leave a stone in its place (see p 272).

Another form of cairn-building is that used by an earthworm, the common and widely-spread *Lumbricus terrestris*. Where the soil is gravel, or where gravel has been used for paths, the pebbles tend to accumulate in heaps six to twelve inches across at the base and two to four inches high. It looks as if gushing rainwater has swept the gravel into heaps. However, there is at the top of each heap a hole of the size an earthworm might make. Darwin referred to this phenomenon in his book *The Formation of Vegetable Mould through the Action of Worms*. This book was published in 1881, and his note about this particular activity has been largely overlooked.

The best time to watch the worm at work is at night, when it is raining slightly. Use a red lamp, as worms are sensitive to white light. By standing quietly one can hear the chink of pebbles being moved, just as one can, in other places, hear the rustle of dead leaves or of pine needles being drawn into the worms' burrows. With the aid of the red light one can see a worm reach out from the hole in the top of the pile, seize a pebble with its mouth, presumably using suction to hold it, then contract the body and drop the pebble on top of the pile. Worms may move pebbles of up to two ounces in this way.

The reason for the worms' behaviour is not clear. However, in less stony places, earthworms can be seen lying along the ground at night with the tail end anchored in the burrow. While in this position, the tip of the worm's front end can be seen moving over the surface of the ground like the nozzle of a vacuum cleaner. Movements of the rings around the region of its gullet suggest that it is swallowing a soft film of algae from the surface of the earth. It may be that where the surface of the ground is covered with small pebbles the worm must first clear the surface of the soil around it in order to feed without leaving its burrow.

Lizard's bloody eyes

The Horned toad, preferably to be called the Horned lizard, of the deserts of the southwestern United States and Mexico, is about five inches long. Its spiny body, with longer spines ('horns') on the head, makes it look like an extinct reptile in miniature. It is known to squirt blood from its eyes, to a distance of fifteen inches. Several people, including specialists on reptiles, have testified to seeing this happen. A student had three of these reptiles and all three squirted blood over his hands the first time he handled them. They

Baffled by a beetle

A well-known scientist was walking with his schoolgirl daughter across a moor. The father was a zoologist, the daughter's hobby was natural history. The father had been holding forth at some length as they walked along about the fundamental difference between man and animals, how man had the power to work things out for himself and animals were actuated almost entirely by blind instinct.

Suddenly they saw in front of them a Dung beetle, black, shiny and about an inch long. The beetle was pulling a deer pellet and walking backwards, as is usual with Dung beetles. At that moment its rear end with its hind legs had become entangled in a tassel of dried grass. The scientist pointed out as they watched the beetle struggling to no purpose that there was a small hole in the ground a few inches beyond the beetle which was clearly the destination at which the beetle was aiming, because it was going to bury the deer pellet. But, he pointed out, the beetle is in difficulty and it hasn't the sense to do anything about it. At that moment the beetle released its hold on the pellet, moved forward slightly, searched around and found a pathway through the matted grass to the hole. 'You see' said the scientist 'blind instinct at work. The

294

The main aim of a desert animal should be to preserve every drop of body fluid yet we have the anomaly that the Horned lizard, a typical desert animal, will squirt blood from its eye sockets. The behaviour is sporadic, random and little understood. The only suggestion so far put forward to account for it is that the blood may possibly deter a predator. Even this is something of an anomaly since the lizard's body is so plentifully covered with spines that alone should minimise predation.

beetle had set out to reach that hole and it is determined to do so with or without the pellet.'

As father and daughter watched, the beetle reached the hole, peered down into it, then turned, retraced its steps along the pathway through the grass, once again seized the pellet, and this time walked backwards following the easy path it had found, reached the hole and dropped the pellet in.

The scientist felt sheepish for a moment. Then he said: 'That beetle showed some sense anyway.'

were three males and each was about to cast its skin. Several theories have been put forward: that it is a means of defence, that it is in some way connected with the breeding season and that it is due to a parasite.

The extraordinary thing about this phenomenon is the contradictory nature of the evidence. R. L. Ditmars, a well-known American herpetologist, witnessed this **auto-hemorrhage** only after having handled hundreds of Horned lizards. H. M. Smith, another distinguished zoologist, saw it only after having handled two hundred. Of the lizards seen squirting blood, some were moulting, others not; some were male, others female.

There are also a few snakes that lose blood through the eyes, for no obvious reason. The best we can say for all these known instances of autohemorrhage is that it may be 'an action of secondary use acquired by a relatively few forms' or we can fall back on the safe refuge of the scientist, that 'the significance of this action is not fully understood' — truly an understatement. Nevertheless, autohemorrhage or **reflex bleeding** by certain insects, including ladybirds, deters predators: the blood may clog their mouthparts, or be poisonous.

Morning symphony

It is not only out-of-the-way phenomena like auto-hemorrhage that still puzzle. The dawn chorus, which is widespread, and has been heard by millions and speculated upon by hundreds, still eludes explanation. This chorus occurs more or less throughout the year, but is most noticeable, and most intense, during the breeding season.

To hear the dawn chorus is one of the compensations for being awake as the gloom of night gives way to dawn. At first there is only scattered desultory song, but as more and more birds join in, the chorus swells until the whole countryside is alive with song. The whole chorus may last for two hours, but it is at its peak during the half-hour before it gradually dies away. Then follows a half-hour's silence during which the birds feed. After this there is a shorter period of less voluminous chorus.

There is a fairly orderly sequence in the times the various species begin to sing. This is to some extent determined by the idiosyncracies of each. The European robin starts to sing three minutes after awakening, first stretching and preening, then flying to a song post. The chaffinch, by contrast, takes twenty minutes to begin. In North America the first to start is the Song sparrow, followed by the cardinal, with the House wrens among the last.

The start of the chorus is related to the intensity of the light. It starts earlier on fine mornings, and later when the sky is overcast or when it is raining. In early spring it starts after sunrise, in summer before sunrise.

One theory is that in singing thus each bird lets the others of its species know it is in occupation of a territory; the singing each morning gives fresh notice not to trespass. However, this would hardly seem to necessitate such an outburst. Despite the detailed documentation of the phenomenon, there is no clear reason why it should occur.

The rat's tail

It is an unfortunate fact of life that too often the questions put to zoologists by laymen are precisely those he either cannot answer or can answer only inadequately. Sometimes this is for lack of evidence rather than, as with the dawn chorus, a super-abundance of it. One such question that crops up from time to time is whether it is true that two rats will combine to transport an egg.

The story is an ancient one, the earliest account being in a thirteenth-century Persian manuscript. This story is one that

Rat to the rescue of rat?
The story was told by two ladies who, disturbed by extensive signs of a rat in the garden, set a trap for it. A little while later they went into the garden and found a large rat caught in the jaws of the trap by the tail, near its base.

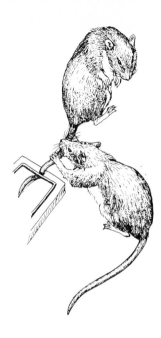

The ensuing moments can be better imagined than described, as one went for a heavy stick to kill the rat and the other went to fetch a pail of water to drown it. Both arrived back from their respective quests simultaneously a few moments later just in time to see a second rat finish gnawing through the imprisoned tail. Then both rats were gone, the one leaving behind the greater part of its tail in the trap.

There is no obvious reason for doubting the story, and, assuming it rests on accurate observation, it furnishes one of those rare examples of intelligent co-operation for which the student of animal behaviour is always looking. It is only one of many anecdotes told about different animals that are trapped. When, however, the story was put before an expert on rat behaviour, he declined to believe a rat would show compassion in this way. He preferred to take the view that the second rat was attacking the trapped rat and happened to choose the tail as the point of attack.

is often derided by zoologists. A rat, so the usual account runs, is seen to lie on its side, take a chicken's egg in its paws and draw it onto its chest. Another rat takes the first one by the tail and drags it along until the two finally disappear out of sight.

Reason, logic and a knowledge of animal behaviour load the scales tremendously against acceptance of this story. It implies a degree of co-operation unusual between animals and it credits rats with an intelligence more appropriate to monkeys, or even to humans. Since rats can readily transport eggs on their own there seems no reason why they should go to this trouble to achieve an end that can be attained single-handed. Above all, the story, on the face of it, sounds too far-fetched.

Nevertheless, there have been scores of people in countries as far apart as New Zealand, the United States and the United Kingdom that have claimed to have seen it. When such people are interrogated their testimony has the ring of truth and their story is hard to pull apart. Moreover, rats often steal eggs from positions where some unusual method must have enabled them to make successful transportation without breaking the eggs. Therefore, the only thing to do, apart from rejecting the story out of hand, is to see how far known facts of rat behaviour can bridge the gap between the anecdotal evidence and the professional zoologists' scepticism.

A rat will transport a chicken's egg as far as a hundred yards using one of several methods. It may carry it in its mouth, in the forepaws with the chin assisting, walking on the hindlegs with the tail pressed on the ground giving additional support, or by bowling it along the ground with the forepaws. One more method has been observed though this is rare. The rat lies on its back, takes the egg between its hindlegs, moves it across its stomach, where it is gripped between the forelegs, and thence pushes it beyond its recumbent head. It then gets onto its feet, walks along, lies down and repeats the manoeuvre. One rat was seen to transport an egg in this fashion for a measured distance of four yards. Had another come on the scene and dragged this rat by the tail while the egg was clasped in its legs and resting on its chest, the ancient story would have been vindicated.

A recent study was made of play in young rats and the results were published in the *Journal of Zoology* (1975). In the social play between juvenile Brown rats there are twelve separate elements. Those relevant here are mouthing, licking each other's fur, biting and pulling the fur, lying flat on the back, orientating sideways to each other, and single biting movements. We know play movements can emerge in adulthood and it needs little more than what is described in the juvenile's play for one rat to pull another by its tail. The gap is narrowing!

Misleading monsters

One powerful argument against acceptance of the story of the two rats is that untrained observers, or even trained observers off their guard, may be deluded by what they see. They therefore weave into their account more than they have actually seen.

There spring to mind two authenticated aspects of animal locomotion that may have led to unravelled mystery. One is the tendency of large aquatic animals to swim in line. The other is **porpoising.** Tunny or tuna are known occasionally to swim in line at the surface with only the dorsal fin clear of the water. There is no known reason for this, although there are plenty of observations of land animals walking in line. Sea-serpent stories often include drawings showing a line of fins and little more. If large sharks or Killer whales ever

Otters swimming in line at the surface have been responsible for some alleged lake monsters.

298

'Who *can* be throwing that stick in the air,' she said to her husband as they trudged up the mountain track in Italy. Clear above the bushes on the steep slope to their right there appeared at intervals what looked like a black stick in the air. It fell into the bushes then appeared again lower down, coming rapidly nearer until it sailed over their heads — a five-foot snake with white belly. They told their story to several experts on snakes, who were frankly disbelieving, though courteous.

A popular belief among Africans in southern Africa is that snakes have phenomenal powers of leaping. A South African citrus farmer once declared that a Puff adder leapt right over one of his orange trees and landed at his feet. But those responsible for snakes in zoos or on snake farms point out that these reptiles never leap the walls of their enclosures.

An Englishman who served in south-east Asia during World War II told the story of a snake progressing in vertical leaps along a road before him, its body coiling as in a corkscrew.

There are other similar stories from other parts of the world. The hamadryad of south-east Asia is generally agreed to progress at times by leaps of three to four feet into the air.

From the viewpoint of muscular output leaping snakes should be no more impossible than porpoises and dolphins leaping 18ft into the air from a standing start, using only a flick of the tail.

After all, an earthworm was once seen to coil into a corkscrew and straighten out to leap several inches into the air.

follow the example of the tuna this could explain these particular sea-serpents.

'Porpoising' is the name given to the action in which porpoises or dolphins proceed at the surface by an undulating movement in which at one moment they are half out of the water and the next moment fully submerged. Seals, sealions and otters also use this method of progression.

William J. Long, the distinguished American naturalist, used to spend his vacations among the lakes of southern Canada, in what he called 'The Wilderness'. He has left in his diary an account of what appeared to be a traditional sea-serpent in one of the lakes. He could hardly believe his eyes as he watched what appeared to be the coils of a huge snake-like animal writhing at the surface. It was a line of otters porpoising. This could well be the explanation of many of the lake monsters reported from various parts of the world.

Good Samaritans

Many of the unsolved problems, like the story of the two rats and the possible existence of lake monsters, lie in a sort of no-man's land between the scientific world and that of the non-scientist. Some of these problems are less concerned with physical aspects of behaviour and more with the emotions and the mental processes. The layman asks the questions, speculates, and often has solid opinions of his own. At least he finds these attractive to contemplate. The scientist usually prefers to push them to one side, refusing to discuss them for want of sufficient data.

One question not infrequently asked is whether any animals are capable of compassion (see p 236). This contrasts starkly with the main areas of zoological research, many of which have been touched upon in previous chapters, in which the dominating features are aggression and predation, inflicting injury and death. If we are honest it has to be admitted that much of human behaviour lies in these same areas and that compassion plays a secondary role.

In the days of big game hunting in Africa, which ended barely more than a half-century ago, those whose main objective was to slaughter nevertheless recorded instances of wounded elephants being succoured by their fellows. An elephant wounded would be helped away by two other elephants ranging themselves on either side of it and supporting it. One group of hunters, having shot an elephant dead, reported that it was soon surrounded by the rest of the herd who, failing to raise their comrade onto its feet, proceeded to drag it through the bush. The hunters, their curiosity aroused, followed and saw the elephants drag the carcase all through the night. The sequel is not recorded.

Instances are known of two otters assisting an injured comrade in a similar manner and of a duck winged by gunshot being assisted in flight by two other ducks ranged on either side of it. Assistance in flight has more commonly been noted where young birds are rescued by one or both parents, which belongs more to the realm of parental or maternal care. Unfortunately, as we have noted previously, our knowledge of these incidents is too often based upon incomplete, and therefore, suspect, observation.

This is a field in which over-enthusiasm can lead the observer astray. In one instance, for example, an enthusiastic but not too competent naturalist watched two crabs that had landed on their backs waving claws and legs in an endeavour to right themselves. Being close together their limbs soon touched, their claws closed on each other's limbs, and soon the crabs were restored to their natural position by what looked like mutual help. The naturalist in question quoted this as an example of compassion. Looked at

Animal co-operatives
There are many stories of rats combining to move heavy objects, such as large wooden beams, even concrete beams, lying on the ground. These have been found moved in circumstances which allow no possibility of their having been moved by humans or other large animals, for instance, in an empty but rat-infested building.

Such feats seem impossible until we recall how ants will apparently co-operate to move heavy loads to their nests. This seeming co-operation has been studied in ants and other insects, and the process seems to take place in three stages. A number of ants will gather around a caterpillar and they will pull in all directions, in complete disorder. Nothing happens to the caterpillar and some of the ants wander away. The others persevere until at last they happen to have moved the caterpillar into the most advantageous position for dragging it, and at the same time all the ants working at it happen to be pulling or pushing in the same direction at the same time. The successful ants continue, at least for a while, and give the impression of working in orderly co-operation.

Another factor that enters into this and makes the stories of the rats plausible is that small animals seem to possess strength out of all proportion to their size. All that is needed then is plenty of time and perseverence and rats have plenty of both.

Rooks and crows sometimes gather around one of their fellows which looks dejected. After a period of lively cawing, all set upon it and peck it to death. This is sometimes called mercy killing, supposedly of a moribund congener.

It is usual to say that talking parrots have no idea of the meaning of the words they use. There is, however, definite evidence that a top class parrot can associate people or situations with particular words or sounds.

One parrot made the sound of the bubbling of water whenever its water pot was replenished. It would do the same at the sight of somebody carrying a jug or a bottle, or even when that person himself drank from a tumbler. One morning the parrot persisted in making the sound of bubbling water interminably, until its owner, curious to know why, walked over to the parrot's cage only to find the water pot as dry as a bone. Someone had forgotten to fill it since the day before, and subsequent events left no doubt the parrot was extremely thirsty. (Compare **vacuum activity** p22.)

The same parrot had a routine. Last thing at night a cloth was thrown over its cage. The owner would say 'Good night' and this was repeated by the parrot.

One morning the parrot was saying 'Good night' so persistently that at last, after an hour or so, it dawned on the preoccupied owner that there must be something amiss, or at least something unusual. He found the parrot hanging head-down from its perch, peering under the edge of the cloth and saying 'Good night'. The parrot clearly associated these words with the movement of the cloth cover in relation to its cage and was asking, in the only way it could, for the cage to be uncovered.

This parrot at least showed the same kind of association between words and objects or actions as the average child does when it first begins to master the use of words.

dispassionately, this is no example of Good Samaritan behaviour but sheer coincidence.

A better example, from among the higher animals as might be expected, comes from the testimony of a competent zoologist who was present when his companion shot two rats, one of which had the tail of the other in its mouth, as they were scurrying away. When the carcases were examined the second rat was found to be totally blind. This gives credence to a long series of stories that have for years been classified with the story of two rats transporting an egg. These are stories of two rats walking side by side and each holding one end of a straw or a stick in their mouths. One rat is blind, and the other is supposed to be guiding it.

Another story, typical of its kind, is of a sheep on a wild moor that was left behind when the flock panicked as a man approached. As the flock jostled through a gap in the stone wall surrounding the field one of the sheep returned to the straggler and nudged it towards the gap. Curious to find out why, the eye-witness, a competent observer, followed hard on the two and found that the straggler left behind by the flock was totally blind.

Kindness through cruelty

The significance of such episodes is perhaps better evaluated by a comparison between mammals and birds. The more convincing instances of what look to be compassion are of mammals which are, especially in their mental capacities, more advanced than birds. In the latter there have been recorded more commonly acts that look like mercy killings or, as they have been called for many years, **compassionate killings.**

Children, even some adults, tend to be merciless towards fellow human beings possessed of some outstanding physical defect, to shun them, deride them, even do them injury. The same trait can sometimes emerge in animals, both mammals and birds. A rabbit with a leg deformed was clearly seen to be ostracized by the rest of the colony and always fed apart from them. Similar instances have been noted in birds. In some species of birds there occurs what appears to be a ritualization of this. In Britain, 'rooks' parliaments' are part of the natural history lore.

Rooks, members of the crow family which, it is generally conceded, includes some of the more intelligent of birds, are gregarious, even social to a degree. Congregations of hundreds of rooks on the ground in a field have been seen, with one dejected, dishevelled individual at the centre. There is much cawing, as if the birds were debating, and in the end first one rook, then another, attacks the dejected one, pecking it to death.

The early interpretation of this was that it was a sort of court of justice in which rooks had gathered to try one of their number who had been guilty of some misdemeanour. Later, it was assumed that the dejected one was diseased and that the rest of the colony were executing it, so putting the individual out of its misery, and at the same time protecting the rest of the colony from a spread of disease.

What is intelligence?

We can argue another way in favour of the possibility that higher animals sometimes show compassion. If there is any truth in the theory of evolution, then much of human behaviour must have its roots in the rest of the animal kingdom. Therefore, if man is capable of compassion it should not be surprising if we find the capability for this emotion somewhere else in the animal kingdom. We ought also to expect that somewhere in the animal kingdom there is true intelligence, an ability to reason, to think, strictly comparable to the capabilities of humans although at a lower level. It is surely only human conceit that leads us to suppose that we have a total monopoly of these capabilities.

Dreaming and recalling

We know people dream just as we know people think, because we can use words to express ideas and convey them one to the other. Dreaming can be detected by the encephalogram (see p 174), but that instrument would be useless if the person being tested could not confirm what had been happening by using speech, which is denied to dogs and animals generally. A sleeping dog will occasionally give a yelp or a muffled bark while in a deep sleep. It is possible also to produce these same effects experimentally by using fine electrodes to stimulate points on the brain of a sleeping or anaesthetised dog. So any manifestations, by voice or body movements, on the part of a sleeping dog, could be no more than the brain being stimulated at certain points by natural processes inside its head.

In extreme cases, this objective explanation can be unconvincing. A dog which is taken into the woods each day for exercise, and which has a wonderful time chasing rabbits may even, in a place where there are no rabbits, go through

Interfering with young animals

We say frequently and without thinking, that if you find a deer fawn and stroke it the mother, detecting human scent on it, will abandon it. People even apply this idea to touching eggs or nestlings, although birds have relatively little sense of smell.

One year, while hay-making, the machine a farm hand was driving severed the hind leg of a roe fawn lying in the grass. Filled with compassion, the farm hand gathered it up and took it home, where it was nursed back to health and lived for years with only three legs. The brother of the injured deer was lying nearby when the accident occurred. To ensure its safety the farm hand then carried it across the field to where the mother could be seen. Approaching fairly closely, he set the fawn down and retreated. The doe called, the fawn answered and ran over to her, and the two retired out of sight into the thickets.

After all, why should the scent of human hands be sufficient to inhibit one of the strongest instincts, the maternal instinct?

Either this popular idea is erroneous or there is some other explanation. For example, deer readily suffer from shock, as when they are netted, often succumbing to it. A doe watching hidden while a group of people handle or molest her baby might suffer a trauma that inhibits the maternal instinct. But that is somewhat different from what we usually say about it. And so far as birds are concerned, disturbance of a nest and its surroundings will upset the parents more certainly than any amount of handling the eggs or the young ones.

302

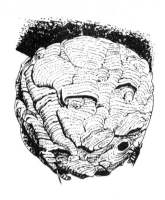

Animal artists
This paper wasp's nest is one of many examples of structures created by animals that have for us an aesthetic appeal. The nests of birds afford many instances as well as nests of insects and the webs of spiders. We even speak of those that make them as animal artists, a term that falls short of true art if only because there is no individuality.

Animal behaviour at the instinctive level is programmed by the genetic code. That is, the animal is born with instructions contained in its genes. It is far easier to programme a regular piece of behaviour. By analogy, machine tools can be controlled by instructions on punched tape. Clearly instructions to drill similar holes at fixed intervals, for instance, will be simpler than instructions to drill a variety of holes at differing distances, or even different sizes of holes at the same distances.

So a wasp building a cell or a Caddis worm making its case in which to live can work to simple instructions which are expressed in a form depending on the size and method of movement of its mouthparts or its legs and the size or nature of the materials available. In other words, the appearance of the finished article appeals to us for much the same reason as normal human works of art appeal to us, because they are orderly.

all the excited movements of limbs and the joyous yelps appropriate to the occasion, as if play-acting at chasing rabbits.

The owner gets to know these scenes well. Then, one day, the dog is fast asleep when its legs begin to jerk, its body heaves and it emits joyous yelps. The owner, watching, sees in all this the familiar pattern, somewhat muted, that his dog shows day after day when in the woods. It takes a lot to convince him that his dog is not dreaming of chasing rabbits.

Telepathy

It would also take a lot to convince some dog owners that their animals do not communicate by telepathy. There is the suspicion that horses may do so as well. Col. J. H. Williams, in his book *Bangaloo,* recounts having tried an experiment to test this. While in India he sent his manservant out into the bush with his dog. Before they set off, Williams and his manservant synchronized watches and the manservant was asked to note what the dog did at precisely 11 a.m. At that hour Williams claims he 'willed' the dog to return home. A short while later the dog entered the house wagging his tail and showing all the signs of a joyful reunion. When the manservant returned he reported that just before eleven o'clock the dog had been sniffing about in the undergrowth. At precisely eleven o'clock the dog had suddenly stopped what it was doing, lifted its head, turned round and made off in the direction of the house.

Maurice Mathis, the French anthropologist, records in his book *Vie et Moeurs des Anthropoïdes* (1954) two striking examples which he witnessed in Africa that raise the suspicion of telepathy being used by both gorillas and chimpanzees. One concerned gorillas observed in the rain forest (p 242). The other, witnessed in its entirety by Mathis himself, involved a male and a female chimpanzee in captivity. In this last incident, on which Mathis commented that if he had not witnessed it himself he would not have believed it possible, the two animals apparently decided upon and carried out a plan of action, and it seemed that they conveyed this plan between them, standing face to face, and making no sound or grimace or gesture, which leaves only communication by ultrasonics or telepathy.

The story of Mathis and his two chimpanzees is too good to leave untold. It concerned a male and a female chimpanzee living in two roomy cages which communicated with each other by a pair of heavy iron sliding doors. On this

303

occasion the doors had been slid open so that the chimpanzees could be together for a while. When, however, the time came for the two to be separated into their respective compartments they refused to be parted. The idea was conceived of placing on the floor in the middle of the female's compartment a bowl of her favourite fruit, then shutting the doors when she went to take it. The fruit was placed in position and all was ready. The two apes approached each other and stood together for a while in the manner already described above. Then the male leapt over to one half of the sliding doors, set his back firmly against the edge of it and held on with his hands so that the doors could not be shut. Meanwhile the female jumped into her compartment, picked up the bowl of fruit and brought it back into the male's compartment where both could enjoy it.

There have been a number of occasions in the past few years when zoologists, experienced in their own fields, have suggested the possibility that animals use telepathy for communication. There have been dog owners who have repeated in some form or other Colonel Williams's experiment, apparently with success.

Telepathy is difficult enough to study in human beings, where both the experimenter and the experimentee can discuss together how the experiment should be framed and how to interpret its results. Neither of these things can be done in experiments with animals. Even so, the frontiers of our knowledge of human and animal behaviour are being slowly pushed back.

General Index

306

Index to Common Names

Note: Where no scientific name follows a common name it is because that common name refers either to a whole family or an order of animals. Where the specific name is not known for certain this is indicated by sp. and where the same name can embrace several species this is indicated by spp.

Index to Scientific Names